A SHORT COURSE ON FUNCTIONAL EQUATIONS

THEORY AND DECISION LIBRARY

General Editors: W. Leinfellner and G. Eberlein

Series A: Philosophy and Methodology of the Social Sciences
Editors: W. Leinfellner (University of Nebraska)
G. Eberlein (University of Technology, Munich)

Series B: Mathematical and Statistical Methods
Editor: H. Skala (University of Paderborn)

Series C: Game Theory and Decision Making
Editor: S. Tijs (University of Nijmegen)

Series D: System Theory, Knowledge Engineering and Problem Solving
Editor: W. Janko (University of Vienna)

SERIES B: MATHEMATICAL AND STATISTICAL METHODS

Editor: H. Skala (Paderborn)

Editorial Board

J. Aczel (Waterloo), G. Bamberg (Augsburg), W. Eichhorn (Karlsruhe),
P. Fishburn (New Jersey), D. Fraser (Toronto), B. Fuchssteiner (Paderborn),
W. Janko (Vienna), P. de Jong (Vancouver), M. Machina (San Diego),
A. Rapoport (Toronto), M. Richter (Aachen), D. Sprott (Waterloo),
P. Suppes (Stanford), H. Theil (Florida), E. Trillas (Madrid), L. Zadeh (Berkeley).

Scope

The series focuses on the application of methods and ideas of logic, mathematics and statistics to the social sciences. In particular, formal treatment of social phenomena, the analysis of decision making, information theory and problems of inference will be central themes of this part of the library. Besides theoretical results, empirical investigations and the testing of theoretical models of real world problems will be subjects of interest. In addition to emphasizing interdisciplinary communication, the series will seek to support the rapid dissemination of recent results.

J. ACZÉL

Centre for Information Theory,
University of Waterloo, Ontario, Canada

A SHORT COURSE ON FUNCTIONAL EQUATIONS

Based Upon Recent Applications
to the Social and Behavioral Sciences

D. REIDEL PUBLISHING COMPANY

A MEMBER OF THE KLUWER ACADEMIC PUBLISHERS GROUP

DORDRECHT / BOSTON / LANCASTER / TOKYO

Library of Congress Cataloging in Publication Data

Aczél, J.

A short course on functional equations.

(Theory and decision library. Series B, Mathematical and statistical methods)

Bibliography: p.

Includes indexes.

1. Economics, Mathematical. 2. Functional equations. 3. Social sciences—Mathematics. I. Title. II. Series.

HB135.A28   1987   515′.25   86–26131

ISBN 90–277–2376–1

ISBN 90–277–2377–X (pbk.)

Published by D. Reidel Publishing Company,
P.O. Box 17, 3300 AA Dordrecht, Holland.

Sold and distributed in the U.S.A. and Canada
by Kluwer Academic Publishers,
101 Philip Drive, Assinippi Park, Norwell, MA 02061, U.S.A.

In all other countries, sold and distributed
by Kluwer Academic Publishers Group,
P.O. Box 322, 3300 AH Dordrecht, Holland.

TABLE OF CONTENTS

Introduction

Recently I taught short courses on functional equations at several universities (Barcelona, Bern, Graz, Hamburg, Milan, Waterloo). My aim was to introduce the most important equations and methods of solution through actual (not artificial) applications which were recent and with which I had something to do. Most of them happened to be related to the social or behavioral sciences. All were originally answers to questions posed by specialists in the respective applied fields. Here I give a somewhat extended version of these lectures, with more recent results and applications included.

As previous knowledge just the basic facts of calculus and algebra are supposed. Parts where somewhat more (measure theory) is needed and sketches of lengthier calculations are set in fine print.

I am grateful to Drs. J. Baker (Waterloo, Ont.), W. Förg-Rob (Innsbruck, Austria) and C. Wagner (Knoxville, Tenn.) for critical remarks and to Mrs. Brenda Law for careful computer-typing of the manuscript (in several versions).

A note on numbering of statements and references: The numbering of Lemmata, Propositions, Theorems, Corollaries and (separately) formulae starts anew in each section. If quoted in another section, the section number is added, e.g. (2.10) or Theorem 1.2. References are quoted by the last names of the authors and the last two digits of the year, e.g. Daróczy-Losonczi [67].

1

1. An aggregation theorem for allocation problems. Cauchy equation for single- and multiplace functions. Two extension theorems.

.

A certain amount s of quantifiable goods (raw materials, energy, money for, say, industrial, agricultural or scientific projects, grants if you wish, etc.) is to be allocated (completely) to m projects. For this purpose (what else?) a committee of n assessors is formed. The i-th assessor recommends that the amount x_{ij} should be allocated to the j-th project. If he (she) can count, then $\sum_{j=1}^{m} x_{ij} = s$. Now the recommendations should be aggregated (synthesized) into a consensus allocation by a chairman or external authority. Again the allocated consensus amounts should *add up to* s. It is supposed (this is a restriction) that the aggregated allocation for the j-th project depends only on the recommended allocations to *that* project, that both the individual and the aggregated allocations be *nonnegative* (!) and, if all assessors recommend rejection, then the consensus allocation to that project should be 0 (*'consensus on rejection'*). If we write the individual allocations for the j-th project as a vector $\mathbf{x}_j$ (Figure 1), then these conditions mean the following ($\mathbb{R}_+ =$

Projects Assessors	1	2	$\cdots$	j	$\cdots$	m	sums
1	x_{11}	x_{12}	$\cdots$	x_{1j}	$\cdots$	x_{1m}	s
2	x_{21}	x_{22}	$\cdots$	x_{2j}	$\cdots$	x_{2m}	s
$\cdots$	$\cdots$	$\cdots$	$\cdots$	$\cdots$	$\cdots$	$\cdots$	$\cdots$
i	x_{i1}	x_{i2}	$\cdots$	x_{ij}	$\cdots$	x_{im}	s
$\cdots$	$\cdots$	$\cdots$	$\cdots$	$\cdots$	$\cdots$	$\cdots$	$\cdots$
n	x_{n1}	x_{n2}	$\cdots$	x_{nj}	$\cdots$	x_{nm}	s
column vectors	$\mathbf{x}_1$	$\mathbf{x}_2$	$\cdots$	$\mathbf{x}_j$	$\cdots$	$\mathbf{x}_m$	$s = (s,...,s)$
aggregates	$f_1(\mathbf{x}_1)$	$f_2(\mathbf{x}_2)$	$\cdots$	$f_j(\mathbf{x}_j)$	$\cdots$	$f_m(\mathbf{x}_m)$	s

Figure 1

$= \{x \mid x \geq 0\}$; when a scalar t stands in place of a vector, $t = (t,...,t)$ is meant):

$$f_j : [0,s]^n \longrightarrow \mathbb{R}_+ \quad (j=1,2,...,m) \tag{1}$$

$$\sum_{j=1}^{m} \mathbf{x}_j = s \Rightarrow \sum_{j=1}^{m} f_j(\mathbf{x}_j) = s \tag{2}$$

and

$$f_j(\mathbf{0}) = 0 \quad (j=1,2,...,m) \ . \tag{3}$$

We emphasize (cf. Aczél-Wagner [81] and Aczél-Kannappan-Ng-Wagner [83]) that both s and m are held fixed (which makes it possible that, for another competition for a different amount and/or with a different number of projects, a different - or the same - 'chairman' may legislate different aggregating functions f_j). - If $m = 2$, the problem of determining f_1, f_2 is trivial: equation (2) becomes

$$f_1(s-\mathbf{x}) + f_2(\mathbf{x}) = s$$

so, in view of (1) and (3), all we have to do is to *choose an f_1 satisfying*

$$f_1(\mathbf{0}) = 0 \; , \;\; f_1(s) = s \; ,$$

$$0 \leq f_1(\mathbf{x}) \leq s \;\; \textit{for all} \;\; \mathbf{x} \in [0,s]^n$$

but *otherwise arbitrary*, and *define* f_2 by

$$f_2(\mathbf{x}) = s - f_1(s-\mathbf{x}) \;\; \textit{for all} \;\; \mathbf{x} \in [0,s]^n \;\; .$$

So from now on we will suppose $m > 2$ (fixed). Substituting into (2) first $\mathbf{x}_1 = s$, $\mathbf{x}_2 = ... = \mathbf{x}_m = \mathbf{0}$ we get, in view of (3), $f_1(s) = s$ and, since the subscript 1 has no privileged role,

$$f_j(s) = s \;\; \text{for all} \;\; j=1,2,...,m \tag{4}$$

('*consensus on overwhelming merit*'). Now we substitute into (2) $\mathbf{x}_1 = \mathbf{z}$, $\mathbf{x}_3 = s-\mathbf{z}$, $\mathbf{x}_2 = \mathbf{x}_4 = ... = \mathbf{x}_m = \mathbf{0}$:

$$f_1(\mathbf{z}) = s - f_3(s-\mathbf{z}) \;\; \text{for all} \;\; \mathbf{z} \in [0,s]^n \;\; .$$

Finally, substitution of $\mathbf{x}_1 = \mathbf{x}$, $\mathbf{x}_2 = \mathbf{y}$, $\mathbf{x}_3 = s-\mathbf{x}-\mathbf{y}$, $\mathbf{x}_4 = ... = \mathbf{x}_m = \mathbf{0}$ (note that we have $m \geq 3$) gives

$$f_1(\mathbf{x}) + f_2(\mathbf{y}) = s - f_3(s-\mathbf{x}-\mathbf{y}) = f_1(\mathbf{x}+\mathbf{y})$$

$$\textit{whenever} \;\; \mathbf{x},\mathbf{y},\mathbf{x}+\mathbf{y} \;\; \textit{are all in} \;\; [0,s]^n \;\; . \tag{5}$$

This is a *functional equation*, that is, an equation where the unknowns are functions. Specifically, it is a (special) *Pexider equation* for multiplace (vector-scalar) functions, (cf. Aczél [66]). Under very weak conditions every Pexider equation can be reduced to a *Cauchy equation* $f(x+y) = f(x) + f(y)$, as we will see later. But in this case the reduction is particularly easy: Put into (5) $\mathbf{x} = \mathbf{0}$ and get, in view of (3), $f_2(\mathbf{y}) = f_1(\mathbf{y})$ for all $\mathbf{y} \in [0,s]^n$. Again, the subscripts 1, 2 have no privileged role, so

$$f_1 = f_2 = \ldots = f_m = f \quad \text{on} \quad [0,s]^n$$

and (5) reduces to the *n-place restricted Cauchy equation*

$$f(x+y) = f(x) + f(y) \quad \text{whenever} \quad x,y,x+y \in [0,s]^n \ . \quad (6)$$

$(x, y, x+y \in [0,s]^m$ is the *restriction*).

It is easy to split (6) into single-place Cauchy equations (cf. Aczél [66]): The equation (6) can readily be generalized (by induction) to

$$f(x_1+x_2+\ldots+x_p) = f(x_1) + f(x_2) +\ldots+ f(x_p)$$

whenever

$$x_1,x_2, \ldots ,x_p,x_1+x_2+\ldots+x_p \in [0,s]^n; \quad p=2,3,\ldots \ .$$

So

$$f(x) = f(x_1,x_2,\ldots,x_n) = f(x_1,0,\ldots,0)$$
$$+ f(0,x_2,0,\ldots,0) +\ldots+ f(0,0,\ldots,0,x_n) \ ,$$
$$(x_1,x_2,\ldots,x_n \in [0,s]) \ .$$

If we write

$$\phi_i(x_i) = f(0,\ldots,0,x_i,0,\ldots,0) \quad (i=1,2,\ldots,n) \qquad (7)$$

(the variable x_i on the *i*-th place), then *we have*

$$f(x) = f(x_1,x_2,\ldots,x_n) = \sum_{i=1}^{n} \phi_i(x_i)$$

$$(x_i \in [0,s]; \ i=1,2,\ldots,n) \qquad (8)$$

where, because of (6) and (7), *every* ϕ_i *satisfies* the restricted Cauchy equation

$$\phi_i(x+y) = \phi_i(x) + \phi_i(y)$$

$$(i=1,2,...,n;\ x,y,x+y \in [0,s]) \quad . \tag{9}$$

In our case, as a consequence of (1), we have

$$\phi_i(x) \geq 0 \text{ for all } i=1,2,...,n \text{ and for all } x \in [0,s] \quad .$$

This inequality and equation (9) imply that the ϕ_i $(i=1,2,...,n)$ are *increasing* (in the broader sense, permitting intervals where ϕ is constant): If $0 \leq x < x' \leq s$, then

$$\phi_i(x') = \phi_i(x) + \phi_i(x'-x) \geq \phi_i(x)$$

$(0 < x' - x \leq s)$. - While we will soon give the *general* solution of (9), we solve it here under this condition of monotonicity (for the sake of brevity we omit the subscript i).

Again, just as for (6), equation (9) implies by induction

$$\phi(x_1+x_2+...+x_p) = \phi(x_1) + \phi(x_2) +...+ \phi(x_p)$$

$$\text{if } x_1,x_2,...,x_p,x_1+x_2+...+x_p \in [0,s]\ ,p = 1,2,... \tag{10}$$

(trivially true for $p = 1$). Consequently (denoting the set of positive integers by $I\!N$),

$$\phi(px) = p\,\phi(x) \text{ if } px \in [0,s]\ ,\ p \in I\!N \tag{11}$$

and

$$\phi(\frac{y}{q}) = \frac{1}{q}\phi(y) \text{ if } y \in [0,s]\ ,\ q \in I\!N \quad . \tag{12}$$

We write now $\phi(s) = a$, so that (12) and (11) give

$$\phi(\frac{s}{q}) = \frac{1}{q}a$$

and

$$\phi\left(\frac{p}{q}s\right) = \frac{p}{q}\phi(s) = \frac{p}{q}a \quad \text{if } p \leq q ,$$

that is,

$$\phi(\rho s) = \rho a \quad \text{for all } \rho \in \mathbb{Q} \cap [0,1] , \tag{13}$$

($\mathbb{Q}$ is the set of all rational numbers). Let now x be an *arbitrary real* number in $[0,s]$. There exist two sequences of rationals in $[0,1]$, $\{\rho_\ell\}$ and $\{R_\ell\}$, the first *increasing* (in the broader sense), the second *decreasing* to x/s, such that $\rho_\ell s \leq x \leq R_\ell s$ ($\ell = 1,2,\ldots,$). Since ϕ is increasing and by (13), we have

$$\rho_\ell a = \phi(\rho_\ell s) \leq \phi(x) \leq \phi(R_\ell s) = R_\ell a$$

As $\ell \to \infty$ we get (with $\alpha = a/s$)

$$\phi(x) = \frac{x}{s}a = \alpha x \quad \text{for all } x \in [0,s] .$$

Since ϕ is increasing, α *is nonnegative*. Conversely, every

$$\phi(x) = \alpha x \quad (x \in [0,s]) ,$$
where α is an arbitrary nonnegative constant, $\tag{14}$

satisfies

$$\phi(x+y) = \phi(x) + \phi(y)$$
$$\text{whenever } x,y,x+y \in [0,s] \tag{15}$$

and

$$\phi(x) \geq 0 \quad \text{for all } x \in [0,s] . \tag{16}$$

So we have proved the following.

Proposition 1. *The general solution of the Cauchy equation (15) under the assumption of nonnegativity (16) is given by (14).*

Using the subscripts i again and taking (8) into consideration, we get

$$\phi_i(x) = \alpha_i x \quad (\alpha_i \geq 0;\ i=1,2,...,n;\ x \in [0,s])$$

and

$$f(\mathbf{x}) = f(x_1,...,x_n) = \sum_{i=1}^{n} \alpha_i x_i$$

$$(\mathbf{x} \in [0,s]^n) \ .$$

Putting this into (4) gives

$$\sum_{i=1}^{n} \alpha_i = 1 \ . \tag{17}$$

Conversely,

$$f_1(\mathbf{x}) = f_2(\mathbf{x}) =...= f_m(\mathbf{x}) = f(x_1,...,x_n)$$

$$= \sum_{i=1}^{n} \alpha_i x_i \quad (\mathbf{x} \in [0,s]) \tag{18}$$

satisfy (1), (2) and (3) with arbitrary nonnegative constants $\alpha_1,\alpha_2,...,\alpha_n$ satisfying (17).

So we have proved the following.

Theorem 2. *The general solution of the system of equations, implications and inequalities (1), (2), (3) is given for $m > 2$ by (18), where $\alpha_1,...,\alpha_n$ are nonnegative constants satisfying (17), but otherwise arbitrary.*

For the original allocation problem this result means that, for $m > 2$, all aggregating functions are *weighted arithmetic means* and, while the 'chairman' *may assign different 'weights' to different assessors,* he (she) cannot change the weights for *different projects during the same competition.*

Aggregation theorems for allocation problems under even weaker conditions can be found in Aczél-Ng-Wagner [84] and another generalization in Radó-Baker [86].

The same argument which has led from (6) to (8) and (9) can be used to *decompose a homomorphism from a cartesian product* $\underset{i=1}{\overset{n}{\times}} N_i$ *of groupoids with neutral elements into a semigroup S.* (A groupoid is a set N_i with an operation $+ : N_i \times N_i \to N_i$. A neutral element 0 satisfies $0 + x = x + 0 = x$ for all $x \in N_i$. While we have n groupoids $N_1, N_2, ..., N_n$, we will denote, without danger of misunderstanding, a neutral element in each of them by 0 and the operation by $+$ in each of them and also in the semigroup S. A semigroup is an associative groupoid. Note that $N_1, ..., N_n$ and S are not supposed to be commutative.) We have now

$$f(x_1+y_1, x_2+y_2, ..., x_n+y_n) = f(x_1, x_2, ..., x_n)$$
$$+ f(y_1, y_2, ..., y_n) \qquad (19)$$
$$\text{for all } x_i, y_i \in N_i \ (i=1,2,...,m) \ .$$

As before, induction gives

$$f(\sum_{j=1}^{p} x_{1j}, \sum_{j=1}^{p} x_{2j}, \ ... \ , \sum_{j=1}^{p} x_{nj}) = \sum_{j=1}^{p} f(x_{1j}, x_{2j}, ..., x_{nj})$$
$$\text{for all } x_{ij} \in N_i \ (i=1,2,...,n; \ j=1,...,p; \ p=2,3,...)$$

Putting $p = n$ and $x_{ij} = 0$ for $i \neq j$, $x_{ii} = x_i$ $(i=1,2,...,n)$, we get

$$f(x_1, x_2, ..., x_n) = \sum_{i=1}^{n} f(0, ..., 0, x_i, 0, ...0)$$

which, with the notation

$$\phi_i(x_i) = f(0, ..., 0, x_i, 0, ..., 0) \quad (x_i \in N_i; \; i = 1, 2, ...n) \quad , \quad (20)$$

becomes

$$f(x_1, x_2, ..., x_m) = \sum_{i=1}^{n} \phi_i(x_i) \quad (x_i \in N_i; \; i = 1, 2, ..., n) \quad . \quad (21)$$

From (19) and (20) we also have

$$\phi_i(x_i + y_i) = \phi_i(x_i) + \phi_i(y_i)$$

$$\text{for all} \quad x_i, y_i \in N_i \; (i = 1, 2, ..., n) \quad . \quad (22)$$

So we have proved the following result. (For clarity, we put after each set the operation under which it forms a groupoid or semigroup, etc.).

Proposition 3. *If* $(N_i, +)$ *$(i = 1, 2, ..., n)$ are groupoids with neutral elements and* $(S, +)$ *a semigroup, then every homomorphism (cf. (19))* $\phi : \underset{i=1}{\overset{n}{\times}} N_i \longrightarrow S$ *can be decomposed in the form (21) into homomorphisms* $\phi_i : N_i \longrightarrow S$ *$(i = 1, 2, ..., n)$ (cf. (22)).*

The converse, that (21) and (22) imply (19), is true if S *(or at least the image of* $\underset{i=1}{\overset{n}{\times}} N_i$ *under* f*) is commutative.*

For similar (but more difficult) theorems on other structures see Kuczma [72] and Szymiczek [73].

Next we give the *general solution of the (unrestricted) Cauchy equation*

$$\phi(x+y) = \phi(x) + \phi(y) \quad (x,y \in \mathbb{R}) \tag{23}$$

both under some weak regularity assumptions on $\phi : \mathbb{R} \to \mathbb{R}$ and without any such assumption.

Just as (10) and (11) followed from (15), we have here

$$\phi(x_1+x_2+...+x_p) = \phi(x_1) + \phi(x_2) +...+ \phi(x_p)$$
$$(x_1,...,x_p \in \mathbb{R}; \ p=2,3,...) \tag{24}$$

and

$$\phi(mx) = m\phi(x) \text{ for all } x \in \mathbb{R}, \ m \in \mathbb{N} \ . \tag{25}$$

With $x = \dfrac{\ell}{m}t \ (\ell,m \in \mathbb{N})$, that is $mx = \ell t$, and using (25) we obtain

$$m\phi(x) = \phi(mx) = \phi(\ell t) = \ell \phi(t)$$

or

$$\phi(\frac{\ell}{m}t) = \frac{\ell}{m}\phi(t) \text{ for all } \ell,m \in \mathbb{N}, \ t \in \mathbb{R} \ .$$

So, for all real t and all positive rational ρ

$$\phi(\rho t) = \rho\phi(t) \ .$$

But we get from the Cauchy equation (23) with $y = 0$ or $y = -x$

$$\phi(0) = 0 \text{ and } \phi(-x) = -\phi(x) \ , \tag{26}$$

respectively, so that we have

$$\phi(\rho t) = \rho\phi(t) \text{ for all } t \in \mathbb{R}, \ r \in \mathbb{Q} \ .$$

Combined with (24) this yields

$$\phi(\sum_{j=1}^{p} \rho_j t_j) = \sum_{j=1}^{p} \rho_j \phi(t_j)$$

$$(t_j \in I\!R, \ \rho_j \in \mathbb{Q}; \ j=1,...,p) \ . \tag{27}$$

Previously we got $\phi(x) = \alpha x$ as the general monotonic solution of a (restricted) Cauchy equation (with $\alpha \geq 0$ if ϕ was supposed to be increasing in the broader sense). We prove now that solutions which are *not* of this form display a very strange behaviour indeed (Hamel [05]).

Theorem 4. *The graph of every solution $\phi: I\!R \longrightarrow I\!R$*

$$\phi(x+y) = \phi(x) + \phi(y) \ \ (x,y \in I\!R) \ , \tag{23}$$

which is not of the form $\phi(x) = \alpha x$ for all $x \in I\!R$, is everywhere dense in the real plane $I\!R^2$.

Proof. The *graph* of ϕ is the plane set

$$G = \{(x,y) \,|\, x \in I\!R, \ y = \phi(x) \in I\!R\} \ . \tag{28}$$

Now take an $x_1 \neq 0$. If ϕ is *not* of the form $\phi(x) = \alpha x$ for any real constant α, then there exists a nonzero real number x_2 such that

$$\frac{\phi(x_1)}{x_1} \neq \frac{\phi(x_2)}{x_2} \ .$$

(Else let $x_2 = x$ run through the nonzero reals and write $\alpha = f(x_1)/x_1$, in order to get $\phi(x) = \alpha x$ for all $x \neq 0$. Because of (26) this holds also for $x = 0$.) But this gives

$$\begin{vmatrix} x_1 & \phi(x_1) \\ x_2 & \phi(x_2) \end{vmatrix} \neq 0 \ ,$$

so that the vectors $\mathbf{p_1} = (x_1, \phi(x_1))$ and $\mathbf{p_2} = (x_2, \phi(x_2))$ are linearly independent and thus *span* the whole plane $I\!R^2$.

This means that to *any* plane vector **p** there exist real numbers r_1 and r_2 such that

$$\mathbf{p} = r_1\mathbf{p}_1 + r_2\mathbf{p}_2 \ .$$

If we permit only *rational* ρ_1, ρ_2 then, by their appropriate choice, we can get with $\rho_1\mathbf{p}_1 + \rho_2\mathbf{p}_2$ *arbitrarily close* to any given plane vector **p** (since $\mathbb{Q}$ is dense in $\mathbb{R}$ and $\mathbb{Q}^2$ in $\mathbb{R}^2$). Now,

$$\rho_1\mathbf{p}_1 + \rho_2\mathbf{p}_2 = \rho_1(x_1, \phi(x_1)) + \rho_2(x_2, \phi(x_2))$$
$$= (\rho_1 x_1 + \rho_2 x_2, \rho_1\phi(x_1) + \rho_2\phi(x_2))$$
$$= (\rho_1 x_1 + \rho_2 x_2, \phi(\rho_1 x_1 + \rho_2 x_2))$$

(the last equality coming from (27)). Thus

$$G_{12} = \{(x,y) \mid x = \rho_1 x_1 + \rho_2 x_2, y = f(x); \ \rho_1, \rho_2 \in \mathbb{Q}\}$$

is everywhere dense in $\mathbb{R}^2$. Since, see (28),

$$G_{12} \subseteq G \ ,$$

the graph G is so much the more dense in $\mathbb{R}^2$. ◻

Corollary 5. *If a function* $\phi: \mathbb{R} \to \mathbb{R}$ *satisfies the Cauchy equation (23) and is continuous at a point or monotonic or bounded from one side on an interval of positive length, then there exists a real constant* α *such that*

$$\phi(x) = \alpha x \quad for \ all \ real \ x \ . \tag{29}$$

Indeed, see Figure 2, under each of these conditions, the graph of ϕ avoids the shadowed region on that figure. Thus

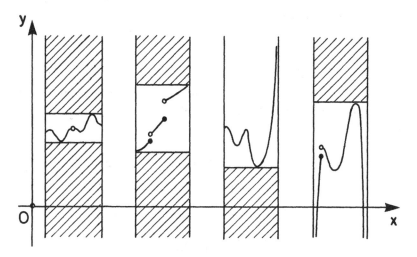

Figure 2

it is not everywhere dense in $\mathbb{R}^2$ and therefore, by Theorem 4, it is of the form (29). - *Conversely, functions of the form (29) with arbitrary constant α's always satisfy (23), but the monotonicity or boundedness assumptions may restrict α. For instance, if ϕ is to be nonnegative (on an interval of nonnegative numbers), then $\alpha \geq 0$.*

If we know (see Steinhaus [20] for a similar result) that, for every set $T \subset \mathbb{R}$ of positive (Lebesgue) measure, the set $T + T$ defined by

$$T+T = \{x+y \mid x \in T, y \in T\}$$

contains an interval of positive length, then we have (cf. e.g. Ostrowski [29]) that *the general solution $\phi : \mathbb{R} \to \mathbb{R}$ of the Cauchy equation (23), bounded from one side on a set of positive measure, is of the form (29).* In particular, *(29) gives the general solution of (3) which is measurable on a set of positive measure.*

Indeed, let ϕ be bounded, say from above, on T:

$$\phi(x) \leq M \quad \text{for all} \quad x \in T .$$

Then, by the Cauchy equation, we have

$$\phi(z) = \phi(x+y) = \phi(x) + \phi(y) \leq 2M \quad \text{for all} \quad z \in T + T \ .$$

Since $T+T$ contains an interval I of positive length, ϕ is bounded from above by $2M$ on I and so, by Corollary 5, is of the form (29). The converse is trivial. - Also functions measurable on a set of positive measure are bounded on a subset of positive measure.

Since we have found the linear functions (29) as the general solutions of the Cauchy equation (23) under very weak assumptions, the question arises whether (29) does not give all solutions of (23) without *any* regularity assumptions on ϕ. The answer is *negative*. Actually, we will determine *all* solutions of the Cauchy equation (23) and see that not all are of the form (29).

For this we need only to know that there exists, in the sense of linear algebra, a basis of $\mathbb{R}$ over $\mathbb{Q}$, the so-called Hamel-basis (Hamel [05]) $B \subset \mathbb{R}$. This means that *every real number x can be expressed in a unique way* (up to 0 coefficients) *in the following form* (the *Hamel expansion* of x):

$$x = \sum_{i=1}^{p} \rho_i h_i,$$

$$where \quad \rho_i \in \mathbb{Q} \quad and \quad h_i \in B \ (i=1,2,...,p) \ . \tag{30}$$

Then *the general solution of the Cauchy equation (23) is constructed by choosing the values of ϕ arbitrarily on B and, if (30) is the Hamel expansion of $x \in \mathbb{R}$, then by defining*

$$\phi(x) = \phi(\sum_{i=1}^{p} \rho_i h_i) = \sum_{i=1}^{p} \rho_i \phi(h_i) \ .$$

That every solution of (23) is of this form follows immediately from (27) and it is easy to see that every function of this form satisfies (23).

It is now clear how solutions of (23) are constructed,

which are not of the form $\phi(x) = \alpha x$: *Let h_1 and h_2 be two* (distinct) *elements of the Hamel basis B. If we assign an arbitrary value $\phi(h_1)$ to h_1 and we also choose $\phi(h_2)$ arbitrarily as long as $\phi(h_2) \neq \dfrac{\phi(h_1)}{h_1} h_2$ (and then choose ϕ completely arbitrarily on the rest of B), then this solution can not be of the form $\phi(x) = \alpha x$.*

In the allocation problem which has served as our point of departure, the *domain* of the Cauchy equation (15) was *restricted* by the condition $x, y, x+y \in [0,s]$. In Proposition 1 we have solved this equation under the nonnegativity assumption appropriate to our problem. We have now the general solution of the *unrestricted* Cauchy equation (23) under much weaker assumptions (and also without any regularity assumptions). We can deduce the corresponding solutions of (15) with aid of *extension theorems*.

We will deal with a somewhat more general situation and prove the following.

Theorem 6A. *Let I be a proper interval of nonnegative numbers (open, half open or closed), with 0 on its boundary (that is, $[0,s],]0,s], [0,s[$ or $]0,s[, s > 0$; also $s = \infty$ permitted). If $\phi : I \rightarrow \mathbb{R}$ satisfies the 'Cauchy equation on a triangle'*

$$\phi(x+y) = \phi(x) + \phi(y) \quad whenever \quad x, y, x+y \in I \ , \quad (31)$$

then there exists a unique function $\psi : \mathbb{R}_+ \rightarrow \mathbb{R}$ which satisfies the Cauchy equation on the nonnegative quarter plane

$$\psi(x+y) = \psi(x) + \psi(y) \quad for \ all \quad x, y \in \mathbb{R}_+ \quad (32A)$$

and is an extension of ϕ, that is,

$$\psi(x) = \phi(x) \quad for\ all \quad x \in I \ . \tag{33}$$

Of course, a similar theorem holds for intervals of nonpositive numbers. Also the following is true (Aczél [83]).

Theorem 6B. *Let $I \subset \mathbb{R}$ be a proper interval (open, half open or closed), with 0 in its interior. If $\phi : I \longrightarrow \mathbb{R}$ satisfies the 'Cauchy equation on a hexagon'*

$$\phi(x+y) = \phi(x) + \phi(y) \quad whenever \quad x,y \in I \ , \tag{31}$$

then there exists a unique function $\psi : \mathbb{R} \longrightarrow \mathbb{R}$ which satisfies the Cauchy equation on $\mathbb{R}^2$

$$\psi(x+y) = \psi(x) + \psi(y) \quad for\ all \quad x,y \in \mathbb{R} \ , \tag{32B}$$

and is an extension of ϕ, that is,

$$\psi(x) = \phi(x) \quad for\ all \quad x \in I \ . \tag{33}$$

The *proof* is the same in both cases (we will call them A and B). Let $x \in \mathbb{R}_+$ or $x \in \mathbb{R}$ be arbitrary in the cases A and B, respectively. Then there exists a positive integer n such that $x/n \in I$ (except if $x = 0$ and $0 \notin I$ in case A - then we define $\psi(0) = 0$; as $y = 0$ shows in (32A), this is the only possible choice). We define

$$\psi(x) = n\phi(\frac{x}{n}) \ . \tag{34}$$

We have to show that this definition is *unambiguous*. In the same way as we obtained (12) from (15), we get

$$\phi(\frac{y}{q}) = \frac{1}{q}\phi(y) \quad for\ all \quad y \in I, q \in \mathbb{N}$$

(both for A and for B). So, if also $x/m \in I$, then we have indeed

$$\frac{1}{m}\phi(\frac{x}{n}) = \phi(\frac{x}{mn}) = \frac{1}{n}\phi(\frac{x}{m}),$$

$$\text{that is, } n\phi(\frac{x}{n}) = m\phi(\frac{x}{m}) \ .$$

We get (33) (that ψ is an *extension* of ϕ) simply by putting $n = 1$ into (34). In order to obtain *the Cauchy equation* (32A,B) *on the extended domains*, we choose n so large that $x/n, y/n, (x+y)/n$ are all in I. Then, by (34) and (31),

$$\psi(x+y) = n\phi(\frac{x}{n}+\frac{y}{n})$$

$$= n\phi(\frac{x}{n})+n\phi(\frac{y}{n}) = \psi(x) + \psi(y)$$

for all $x \in \mathbb{R}_+$ in case A (also $\psi(x+0) = \psi(x) + \psi(0)$, of course) and for all $x \in \mathbb{R}$ in case B. As a consequence,

$$\psi(\frac{x}{n}) = \frac{\psi(x)}{n} \tag{35}$$

for all $x \in \mathbb{R}_+$ or $x \in \mathbb{R}$, respectively. - Finally, ψ *is unique.* Indeed, if $\widetilde{\psi}$ also had the properties (33) and (35) then, for arbitrary $x \in \mathbb{R}_+$ or $x \in \mathbb{R}$, choosing $n \in \mathbb{N}$ so that $x/n \in I$, we get

$$\psi(x) = n\phi(\frac{x}{n}) = n\widetilde{\psi}(\frac{x}{n}) = \widetilde{\psi}(x) \ . \qquad \square$$

In case A we have extended (31) only to the nonnegative quarter $\mathbb{R}_+^2$, not to the whole plane $\mathbb{R}^2$. We can further extend (32A) to (32B). We will prove a more general theorem (Aczél-Baker-Djokoviç-Kannappan-Radó 1971; also the above theorems can be generalized to more general algebraic structures, cf. Ng [74], Dhombres-Ger [78]).

Theorem 7. *Let $(G,+)$ and $(H,+)$ be abelian groups and $(S,+)$ a semigroup which generates G. Then every homomorphism $\psi: S \to H$ can be uniquely extended to a homomorphism $G \to H$. In particular, for every solution $\psi: \mathbb{R}_+ \to \mathbb{R}$ of (32A) there exists a unique extension $\widetilde{\psi}: \mathbb{R} \to \mathbb{R}$ which satisfies (32B).*

Proof. The semigroup $(S,+)$ *generates* the abelian group $(G,+)$ if, for every $t \in G$ there exist $x, y \in S$ such that $t = x-y$. This is clearly true when $S = \mathbb{R}_+$, $G = \mathbb{R}$ and $+$ is the ordinary addition of reals.

By supposition, $\psi: S \to G$ is a homomorphism:

$$\psi(x+y) = \psi(x) + \psi(y) \quad \text{for all} \quad x,y \in S \ . \qquad (36)$$

(Note that the operations in the groups G and H have both been denoted by $+$. This was done for the sake of (32A) and should not lead to any misunderstanding.) Take an arbitrary $t = x-y \in G$ $(x,y \in S)$ and define f by

$$f(t) = f(x-y) = \psi(x) - \psi(y) \ . \qquad (37)$$

Again, this definition is *unambiguous*: If $t = x-y = u-v$ $(x,y,u,v \in S)$, then $x+v = u+y$ and, by (36),

$$\psi(x) + \psi(v) = \psi(x+v) = \psi(u+y) = \psi(u) + \psi(y) \ .$$

So indeed

$$\psi(x) - \psi(y) = \psi(u) - \psi(v) \ .$$

Also, f is an *extension* of ψ: If $t = x-y \in S$, then $x = t+y$ $(x,t,y \in S)$ and, by (36) and (37),

$$\psi(x) = \psi(t+y) = \psi(t) + \psi(y), \quad \text{that is,}$$

$$f(t) = \psi(x) - \psi(y) = \psi(t) \quad .$$

Further, f is a *homomorphism* of G into H. Indeed, if $t = x-y$ and $w = u-v$ are arbitrary in G $(x,y,u,v \in S)$ then $t+w = x-y+u-v = (x+u)-(y+v)$. So

$$f(t+w) = \psi(x+u)-\psi(y+v)$$

$$= \psi(x)+\psi(u)-\psi(y)-\psi(v)$$

$$= f(t)+f(w) \text{ for all } t,w \in G \quad ,$$

because of (36) and (37). - Finally, f is *unique*. Indeed, if $\tilde{f}:G \longrightarrow H$ were another extension of the homomorphism $\psi: S \longrightarrow H$ then, for all $t = x-y \in G$ $(x,y \in S)$,

$$\tilde{f}(t) = \tilde{f}(x-y) = \tilde{f}(x)-\tilde{f}(y)$$

$$= \psi(x)-\psi(y) = f(x-y) = f(t) \quad . \qquad \square$$

In view of Theorems 6A, 6B, 7 and Corollary 5 we have also the following (cf. Aczél [83]).

Corollary 8. *Let $I \subset \mathbb{R}$ be an interval either containing 0 or having 0 on its boundary. If $\phi:I \longrightarrow \mathbb{R}$ satisfies*

$$\phi(x+y) = \phi(x) + \phi(y) \quad \text{whenever} \quad x,y,x+y \in I \qquad (38)$$

then there exists a unique extension $\psi:\mathbb{R} \longrightarrow \mathbb{R}$ of ϕ which satisfies (32B). If, moreover, ϕ is continuous at a point or monotonic or bounded from one side on a subinterval of I of positive length, then there exists a real constant α such that

$$\phi(x) = \alpha x \quad \text{for all} \quad x \in I \quad . \qquad (39)$$

Conversely, all functions of the form (39) satisfy (38).

Of course, boundedness on a proper interval can again

be replaced by boundedness on a set of positive measure. Also, *the general solution of the restricted Cauchy equation (38) can be obtained by restricting the general* (Hamel) *solution of the unrestricted Cauchy equation (23) to* I.

On the other hand, from Proposition 3 and Corollaries 5 and 8 (or from their proofs) we have the following.

Corollary 9. *Let* $I_i \subseteq \mathbb{R}$ $(i=1,...,n)$ *be proper intervals containing* 0 *(or having* 0 *on their boundary). If* $f:I_1 \times...\times I_n \rightarrow \mathbb{R}$ *is continuous at a point or bounded from one side on a proper* n*-dimensional subinterval of* $I_1 \times...\times I_n$ *and satisfies*

$$f(x_1+y_1,...,x_n+y_n) = f(x_1,...,x_n)+f(y_1,...,y_n)$$

whenever $x_i,y_i,x_i+y_i \in I_i$; $(i=1,...,n)$,   (40)

then there exist real constants $\alpha_1,...,\alpha_n$ *such that*

$$f(x_1,...,x_n) = \sum_{i=1}^{n} \alpha_i x_i \quad for\ all \quad x_i \in I_i (i=1,...,n) \ .$$
   (41)

Conversely, all functions of the form (41) satisfy (40).

This can give results similar to Theorem 2 under weaker conditions.

2. Scale-invariant equal sacrifice in taxation. The linear-affine functional equation. Multiplicative and logarithmic functions.

In the theory of taxation (as in many parts of economics) it is not the amount of, say, income that counts, but its *utility*. If u is the utility function, the utility of the gross income x is $u(x)$ and that of the net income y (after taxes) is $u(y)$ then the (absolute) *sacrifice* under taxation is

$$u(x) - u(y) \ .$$

Under the 'equal sacrifice principle' this should be constant. Of course, all this depends upon the *utility function u*. Whether the sacrifice in general is equal or not, it is reasonable to suppose that it be *scale-invariant* (that is, independent of the monetary unit in which incomes are measured) for the incomes of those who sacrifice equally. In other words

$$u(x) - u(y) = u(x') - u(y') \Rightarrow u(rx) - u(ry)$$

$$= u(rx') - u(ry') \ (x, y, x', y', \ r \in \mathbb{R}_{++}) \tag{1}$$

$(\mathbb{R}_{++} = \{x \mid x > 0\})$. It is quite surprising that *this simple requirement*, together *with rather weak regularity requirements* (that u be *continuous and nonconstant*) *completely determines the possible utility functions*: essentially, they are (affine transforms of) *logarithms or power functions* (cf. Young [86]).

Indeed, the implication (1) clearly means that

22

$$u(rx) - u(ry) = F[u(x)-u(y),r]$$

$$(x,y,r \in \mathbb{R}_{++}) . \tag{2}$$

We determine u by first calculating the function F.

Consider the set

$$I = \{u(x)-u(y) \mid x,y \in \mathbb{R}_{++}\} .$$

Since u is continuous and nonconstant, I is a proper interval containing 0 (for $x = y$) and actually symmetric with respect to 0. Indeed, the image of $\mathbb{R}_{++}$ under u is a proper interval, J, and

$$I = J-J = \{\xi-\eta \mid \xi \in J, \eta \in J\} .$$

In (2) the function F is defined on $I \times \mathbb{R}_{++}$.

Take any s and t such that

$$s,t,s+t \in I .$$

We prove that there exist $x,y,z \in \mathbb{R}_{++}$ such that

$$s = u(x) - u(y), \quad t = u(y) - u(z),$$

$$\text{and so} \quad s+t = u(x) - u(z) , \tag{3}$$

that is, there exist $\xi,\eta,\varsigma \in J$, such that

$$s = \xi-\eta, \quad t = \eta-\varsigma, \quad s+t = \xi-\varsigma .$$

First we choose $s \geq 0$, $t \geq 0$ (and so also $s+t \geq 0$). Then, since $s+t \in I$,

$$s+t = \xi-\varsigma .$$

Since $0 \leq s \leq s+t = \xi-\varsigma$, therefore, if we take $\eta = \xi-s = \varsigma+t$, then $\varsigma \leq \eta \leq \xi$ and, with $\xi \in J$, $\varsigma \in J$, also $\eta \in J$ (J being an interval). So indeed $s = \xi-\eta$ and $t = \eta-\varsigma$.

Another typical situation is $s \leq 0 \leq t$ and, say, $s+t \geq 0$. Then also $s' = -s \geq 0$ is in I (which is symmetric with respect to 0) and so is $t' = s+t \geq 0$ and of course, also $s'+t' = t \geq 0$ is in I. So we have our above first case $s' \geq 0$, $t' \geq 0$, $s'+t' \geq 0$ and thus there exist $\xi',\eta',\varsigma' \in J$ such that

$$-s = s' = \xi'-\eta', \ s+t = t' = \eta'-\varsigma'$$

$$\text{and} \ \ t = s'+t' = \xi'-\varsigma' \ .$$

With $\xi = \eta'$, $\eta = \xi'$, and $\varsigma' = \varsigma$ we have $s = \xi-\eta$, $t = \eta-\varsigma$, $s+t = \xi-\varsigma$ as asserted. One sees immediately how the remaining cases can also be reduced to the first.

But from (2)

$$u(rx) - u(ry) = F[u(x)-u(y),r],$$

$$u(ry) - u(rz) = F[u(y)-u(z),r]$$

and

$$u(rx) - u(rz) = F[u(x)-u(z),r] \ .$$

So

$$F[u(x)-u(z),r] = F[u(x)-u(y),r]$$
$$+ F[u(y)-u(z),r] \ (x,y,z,r \in \mathbb{R}_{++})$$

or, cf. (3),

$$F(s+t,r) = F(s,r) + F(t,r)$$

$$\text{whenever} \ s,t,s+t \in I, \ r \in \mathbb{R}_+ \ .$$

Keeping r fixed for a moment we have the Cauchy equation on a hexagon and applying Corollary 1.8 we get, since by (2) the continuity of u implies that of F in the first (and also in the second) variable,

$$F(t,r) = \alpha(r)t \qquad (4)$$

(the 'constant' α may depend upon r which we had fixed).

If we put (4) back into (2), we obtain

$$u(rx) - u(ry) = \alpha(r)[u(x)-u(y)] \ .$$

Keeping y constant and writing $\beta(r) = u(ry) - \alpha(r)u(y)$, this becomes

$$u(rx) = \alpha(r)u(x) + \beta(r) \quad (x,r \in \mathbb{R}_{++}) \qquad (5)$$

the 'linear-affine functional equation' (so called, because $x \mapsto rx$ is a (homogeneous) linear and $u \mapsto \alpha u + \beta$ an affine (inhomogeneous linear) transformation).

Notice that (5) contains three unknown functions u, α and β (even though we are interested only in determining u), just as we had two in (1.5). It is remarkable that one functional equation can determine several unknown functions. We proceed by reducing the number of unknown functions. For this purpose, first put $x = 1$ into (5):

$$u(r) = \alpha(r)u(1) + \beta(r)$$

and, by subtracting this from (5), get the functional equation

$$v(rx) = \alpha(r)v(x) + v(r) \qquad (6)$$

for the function v defined by

$$v(x) = u(x) - u(1) \quad (x \in \mathbb{R}_{++}) \ . \qquad (7)$$

The equation (6) contains only two unknown functions. In solving (6) we distinguish two cases.

First, if

$$\alpha(r) = 1 \text{ for all positive } r,$$

then (6) goes over into

$$v(rx) = v(r) + v(x) \quad (x, r \in \mathbb{R}_{++}) .\qquad (8)$$

Functions satisfying (8) are called *logarithmic*. This equation can be easily reduced to the Cauchy equation (1.23). Indeed, writing

$$r = e^s, \ x = e^t \ \text{ and } \ v(e^t) = \phi(t) \ (s, t \in \mathbb{R})$$

we get

$$\phi(s+t) = \phi(s) + \phi(t) \text{ for all real } s, t .\qquad (9)$$

So $v(x) = \phi(\log x)$ *is the general solution of (8), where ϕ is an arbitrary solution of (9)* and, by (7), $u(x) = \phi(\log x) + b$, where b is an arbitrary constant. If, as in our case, u is continuous, so is ϕ and we get by Corollary 1.5 $\phi(t) = at$ and *the general continuous solution of (8) is* $v(x) = a \log x$. From (7)

$$u(x) = a \log x + b \quad (x \in \mathbb{R}_{++})\qquad (10)$$

where a and b are constants. Since, by supposition, u is not constant, $a \neq 0$, but otherwise a and b are arbitrary.

The remaining case is when $\alpha(t) \not\equiv 1$. Then there exists an r_0 such that

$$\alpha(r_0) \neq 1 .\qquad (11)$$

Notice that the left hand side of (6) is symmetric in x and r, so the right hand side must be symmetric too:

$$\alpha(r)v(x) + v(r) = \alpha(x)v(r) + v(x) \ .$$

Put into this equation the $r = r_0$ from (11) and write $a = v(r_0)/[\alpha(r_0)-1]$ in order to get

$$v(x) = a[\alpha(x)-1] \quad (x \in \mathbb{R}_{++}) \ , \tag{12}$$

Since u and thus v is not constant, $a \neq 0$. With (12) the equation (6) becomes

$$a\alpha(rx) - a = a\alpha(r)\alpha(x) - a\alpha(r) + a\alpha(r) - a \ ,$$

that is,

$$\alpha(rx) = \alpha(r)\alpha(x) \quad (r,x \in \mathbb{R}_{++}) \ , \tag{13}$$

(only one unknown function left). Functions satisfying (13) are called *multiplicative*. Also this equation can be reduced to the Cauchy equation or to (8). It is clear that we will do the latter by taking logarithms of both sides of (13). For this we have to establish that α is everywhere positive. Actually, $\alpha(x) \equiv 0$ is a solution of (13) but we *exclude* this (since, by (12) and (7), it would cause u to be constant). Then α is *nowhere* 0 because $\alpha(x_0) = 0$ would imply $\alpha(x) = \alpha(\frac{x}{x_0}x_0) = \alpha(\frac{x}{x_0}) = \alpha(\frac{x}{x_0})\alpha(x_0) = 0$ for all positive x. Furthermore,

$$\alpha(x) = \alpha(\sqrt{x} \cdot \sqrt{x}) = \alpha(\sqrt{x})^2 > 0$$

for all positive x ,

as required. So now we can take logarithms of both sides of (13) and get

$$\log \alpha(rx) = \log \alpha(r) + \log \alpha(x) \ ,$$

an equation of the form (8). So *the general not identically zero solution of (13) is given by*

$$\alpha(x) = e^{\phi(\log x)} \quad (x \in \mathbb{R}_{++}) \ ,$$

where ϕ is an arbitrary solution of (9), and, by (12) and (7),

$$u(x) = e^{\phi(\log x)} + b \ .$$

If u is continuous, so is ϕ, and we have again, by Corollary 1.5, $\alpha(x) = e^{c \log x} = x^c$ as the general continuous, not identically zero solution of (13) and

$$u(x) = ax^c + b \quad (x \in \mathbb{R}_{++}) \tag{14}$$

with $a \neq 0$, $c \neq 0$, if u is to be nonconstant. So we have proved the following.

Theorem 1. *The general continuous nonconstant utility functions generating scale-invariant equal sacrifices, that is, the general continuous nonconstant solutions of (1) or (2) are given by (10) and (14) with $a \neq 0$, $c \neq 0$ but the constants a,b,c otherwise arbitrary.*

While $u(x) - u(y)$ is the absolute sacrifice,

$$\frac{u(x)-u(y)}{u(x)} = 1 - \frac{u(y)}{u(x)}$$

is the *relative sacrifice* (one has to suppose here, not unreasonably, that $u(x) \neq 0$) and the requirement that this be *scale-invariant*, means

$$\frac{u(y)}{u(x)} = \frac{u(y')}{u(x')} \Rightarrow \frac{u(ry)}{u(rx)} = \frac{u(ry')}{u(rx')}$$

$$(x,y,x',y',r \in \mathbb{R}_{++}) \tag{15}$$

or

$$\frac{u(ry)}{u(rx)} = G\left[\frac{u(y)}{u(x)}, r\right]$$

$$(x,y,r \in \mathbb{R}_{++}) \ . \tag{16}$$

If (again not unreasonably) we *suppose that* $u(x) > 0$ *for all* $x > 0$ and define

$$U(x) = \log u(x), \ F(t,r) = -\log G(t^{-1},r) \ ,$$

then (15) and (16) are transformed into

$$U(x) - U(y) = U(x') - U(y') \Rightarrow U(rx) - U(ry)$$

$$= U(rx') - U(ry')$$

$$(x,y,x',y',r \in \mathbb{R}_{++})$$

or

$$U(rx) - U(ry) = F[U(x)-U(y),r]$$

$$(x,y,r \in \mathbb{R}_{++}),$$

that is, up to a slight change of notation, into (1) or (2), respectively. So we have the following.

Corollary 2. *The general positive, continuous and nonconstant utility functions generating scale invariant relative sacrifices, that is, the general positive, continuous and nonconstant solutions of (15) or (16) are given by*

$$u(x) = Bx^a \quad (x \in \mathbb{R}_{++}) \tag{17}$$

and by

$$u(x) = Be^{ax^c} \quad (x \in \mathbb{R}_{++}) \tag{18}$$

where $a \neq 0$, $c \neq 0$, $B > 0$, but otherwise a, B, c are arbitrary real constants.

Of course, further restrictions on the utility functions are reasonable, for instance that u be *increasing*. *Then in (10) and (17) $a > 0$, in (14) and (18) either $a > 0$, $c > 0$ or $a < 0$, $c < 0$. In (14), $a < 0$, $c < 0$ has its *disadvantage*: $u(x) < 0$ if $x < (-b/a)^{1/c}$ and* one would think that no amount of positive income should have *negative utility*.

What does all this mean for taxation? First, the last problem may not come up if we consider taxation (and therefore also utility) only from a certain minimal income (say, γ) upward. But then (1) and (2) (and also (5) and (16)) should be considered only for $x,y,x',y' \geq \gamma$, making them *restricted* equations. These can be solved (cf. Stehling [74, 75], Aczél [84]; also section 7 below).

On the other hand, let f be the *taxation function*, that is, $f(x)$ the amount of tax paid on the (gross) income x. Then the net income is

$$y = x - f(x)$$

and, if the taxation is supposed to lead to *equal (absolute) sacrifice $d > 0$*, then

$$u(x) - u[x - f(x)] = d \quad .$$

Now let the utility function u be *continuous and strictly monotonic,* as it turned out to be in Theorem 1. Then it has an inverse u^{-1} and the taxation function f can be expressed in terms of u:

$$f(x) = x - u^{-1}[u(x) - d] \ . \tag{19}$$

Take those u which are *strictly increasing*. For (10) we get

$$f(x) = x - e^{-d/a}x = (1 - e^{-d/a})x$$

$$(a > 0, \ d > 0) \tag{20}$$

while for (14)

$$f(x) = x - (x^c - \frac{d}{a})^{1/c} \ . \tag{21}$$

In this latter case we had either $a > 0$, $c > 0$ or $a < 0$, $c < 0$.

If we want the taxation to be *progressive*, then the function g defined by

$$g(x) = \frac{f(x)}{x}$$

should be *increasing* in the broader sense, which permits also constant g's. For *the taxation function (20) g is indeed constant*. As to (21),

$$g(x) = 1 - (1 - \frac{d}{ax^c})^{1/c} \ . $$

If $a > 0$, $c > 0$, then this g is *strictly decreasing, so the taxation is regressive. If $a < 0$, $c < 0$ then g is increasing, the taxation is progressive* but, as we have seen, the utility function is negative for $x < (-b/a)^{-1/c}$.

Let us look now at *equal relative sacrifice D*. It is reasonable to suppose $0 < D < 1$ (excluding both total and zero taxation). Then

$$\frac{u(x)-u\,[x-f(x)]}{u(x)} = D \in \,]0,1[$$

and, if u is continuous and strictly monotonic,

$$f(x) = x - u^{-1}[u(x)(1-D)] \quad . \tag{22}$$

Now we have to use the utility functions (17) and (18) which generate *scale-invariant relative sacrifices*. For (17)

$$f(x) = [1-(1-D)^{1/a}]x \quad (0 < D < 1,\ a > 0)$$

(the same as (20), with $d = -\log(1-D)$), which gives *constant taxation rate*

$$\frac{f(x)}{x} = 1 - (1-D)^{1/a} \quad .$$

Finally, for (18)

$$f(x) = x - (x^c + \frac{\log(1-D)}{a})^{1/c}$$

(the same as (21), with $d = -\log(1-D)$), which gives the taxation rate

$$g(x) = \frac{f(x)}{x} = 1 - (1+\frac{\log(1-D)}{ax^c})^{1/c} \quad .$$

If $a > 0$, $c > 0$, then g *decreases* strictly since $0 < D < 1$) so *the taxation is regressive*. If, on the other hand, $a < 0$, $c < 0$, let us write $A = -a > 0$ and $C = -c > 0$. So

$$f(x) = x - x[1 - \frac{1}{A}x^C \log(1-D)]^{-1/C}$$

$$(A > 0,\ C > 0,\ 0 < D < 1) \tag{23}$$

and

$$g(x) = \frac{f(x)}{x} = 1 - [1 - \frac{1}{A}x^C \log (1-D)]^{-1/C}$$

is increasing, so the taxation is *progressive*. Since

$$u(x) = Be^{-Ax^{-C}} \quad (A > 0, B > 0, C > 0) \tag{24}$$

(coming from (18) with $A = -a$, $C = -c$) is always positive, from this point of view *the taxation function (23) with the utility function (24) seems to make the most sense* (notice also that, for (24), $\lim_{x \to 0} u(x) = 0$ and $\lim_{x \to \infty} u(x) = B$: the utility of 0 income is 0 and the utility does not grow over all bounds with increasing income). If we restrict ourselves to $x > (A/b)^{1/C}$, then also the utility function

$$u(x) = b - \frac{A}{x^C}$$

makes sense with the taxation function

$$f(x) = x - x(1 + \frac{d}{A}x^C)^{-1/C}$$

$(A > 0, b > 0, C > 0, d > 0)$, coming with $A = -a$, $C = -c$ from (14) and (21), respectively. Of course, this is again *the same as (23)* $(d = -\log (1-D) > 0)$ and then the tax rate satisfies (in both instances)

$$\lim_{x \to 0} g(x) = \lim_{x \to 0} \frac{f(x)}{x} = 0 ,$$

$$\lim_{x \to \infty} g(x) = \lim_{x \to \infty} \frac{f(x)}{x} = 1 ,$$

as presumably one should expect.

(The reason why the taxation function f is the same for equal absolute and for equal relative sacrifices is the following. As we have seen, (15) (scale-invariant equal relative sacrifices) goes over into (1) (scale-invariant equal absolute sacrifices) with

$$U(x) = \log u(x) \tag{25}$$

and, accordingly, we obtained (17) and (18) from (10) and (14) by replacing u by this U. But the equation (22) determining f in the case of equal *relative* taxation can be written as

$$f(x) = x - U^{-1}[U(x) - d] \quad,$$

where U is defined exactly by (25) (and, accordingly, $U^{-1}(z) = u^{-1}(e^z)$ and $d = -\log(1-D)$. This is then indeed the same as (19).)

3. General forms of 'laws of science' without dimensional constants. The case of the same ratio scale for all variables. Generalized homogeneous functions. Inequality measures.

In the previous section changes of scales played an important role. This role is quite fundamental in the natural, behavioral and social sciences.

Dimensional analysis (for instance in physics) is based on the recognition that the laws (formulae) of science should be equations, both sides of which have the same (physical) dimensions — if there are no dimensional constants. This implies mathematically that these laws should remain invariant under a change of scales (we will call these now *'ratio scales'* because we will deal below also with other kinds of scales), that is, (homogeneous) linear transformations of the 'independent' variables generate similar transformations of the function value ('dependent variable'):

$$u(r_1 x_1,...,r_n x_n) = R(r_1,...,r_n)u(x_1,...,x_n)$$

$$(x_i > 0, r_i > 0;\ i=1,...,n;\ R,u : I\!\!R_{++}^n \longrightarrow I\!\!R_{++})\ . \tag{1}$$

Not all scales are ratio scales, for instance temperature scales are not (consider, for instance, conversion from Fahrenheit to centigrade (Celsius) scales), even though this problem is eliminated in physics by taking the 'absolute' (Kelvin) temperature scale or considering only differences in temperature. Some quantities, like temperature, have *'interval scale'*, they change by affine (inhomogeneous linear)

transformations $x_i \mapsto r_i x_i + p_i$ $(x_i, p_i \in \mathbb{R}, r_i \in \mathbb{R}_{++})$.
Strictly speaking also time is on an interval scale, both its
unit and its 'zero' (starting point) can change. This normally
does not come up in physics either, again because only or
mostly differences in time are considered. But in economics,
for instance when forming inflation indicators, not only the
unit but also the starting point ('zero') of counting time is
often changed. The so-called 'logarithmic' scales in psychophysics are really transformed interval scales.

In psychophysics, psychology, economics etc., when the
independent variables have interval scales the dependent
variable may have interval or ratio scales and vice-versa
(Luce [59, 64]). Also other scales are used (we have mentioned already the 'logarithmic' scales but they can easily be
reduced to ratio or interval scales) and sometimes different
kinds of scales occur for different (independent) variables.

Another generalization may be needed. The (linear or
affine) transformations in the different independent variables
may not be chosen independently. For instance, in the price
index $\sum\limits_{j=1}^{m} p_j / \sum\limits_{j=1}^{m} p_{j0}$ the change of (ratio) scale, say to a different currency, has to be the *same* for all prices. In cases of
interval   scales   changed   by   affine   transformations
$x_i \mapsto r_i x_i + p_i$, the $r_i (\in \mathbb{R}_{++})$ may be the same (*same unit*)
or the $p_i (\in \mathbb{R})$ ('*same zero*') or both. One can determine all
laws (functions), where the *dependent variable (function
value) and the independent variables have ratio or interval
scales and the transformations may or may not apply
independently.* We deal here with eight of those twelve
cases where *all independent variables allow the same type of
transformations* (affine or linear, equal or independent; for
more details on others see Aczé l-Roberts-Rosenbaum [86]).

In this section we are first concerned with the case where all independent variables undergo the same linear transformation, resulting in an affine transformation of the dependent variable. In other words, we solve the functional equation

$$u(rx_1,...,rx_n) = R(r)u(x_1,...,x_n) + P(r)$$

or, in vector notation, $\mathbf{x} = (x_1,...,x_n) \in \mathbb{R}_{++}^n$

$$u(r\mathbf{x}) = R(r)u(\mathbf{x}) + P(r)$$

$$(r \in \mathbb{R}_{++}, \; \mathbf{x} \in \mathbb{R}_{++}^n; \; u : \mathbb{R}_{++}^n \to \mathbb{R},$$

$$P : \mathbb{R}_{++} \to \mathbb{R}, \; R : \mathbb{R}_{++} \to \mathbb{R}_{++}), \qquad (2)$$

where the vector $\mathbf{x}$ is multiplied by the scalar r componentwise: $r\mathbf{x} = r(x_1,...,x_n) = (rx_1,...,rx_n)$. As in section 1, when a scalar t stands in place of a vector, $t = (t,...,t)$ is meant. Notice that (2) is a generalization of (2.5) to which it reduces if $n = 1$. We can also solve it in a similar way.

In order to solve (2), first substitute $\mathbf{x} = 1$ into it, then subtract the equation thus obtained from (2) in order to get, as before,

$$v(r\mathbf{x}) = R(r)v(\mathbf{x}) + v(r)$$

$$(r \in \mathbb{R}_{++}, \; \mathbf{x} \in \mathbb{R}_{++}^n) \qquad (3)$$

for

$$v(\mathbf{x}) = u(\mathbf{x}) - u(1) \; . \qquad (4)$$

We again distinguish two cases.

First, if $R(r) \equiv 1$, then (3) goes over into

$$v(r\mathbf{x}) = v(\mathbf{x}) + v(r) \; . \qquad (5)$$

In the particular case $\mathbf{x} = (s,...,s) = s$, this reduces to

$$v(rs) = v(r) + v(s) \quad (r,s \in \mathbb{R}_{++})$$

that is, to (2.8). We have called functions satisfying this equation 'logarithmic' functions. To remind us of this we will denote by L any function which satisfies

$$L(rs) = L(r) + L(s) \quad (r,s \in \mathbb{R}_{++}) \; . \tag{6}$$

So we have now $v(r) = L(r)$. Notice that this is a statement about $v(r) = v(r,...,r)$, not about $v(\mathbf{x}) = v(x_1,...,x_n)$. However, we can write (5) as

$$v(r\mathbf{x}) = v(\mathbf{x}) + L(r)$$

and determine $v(\mathbf{x})$ immediately, for instance in the form

$$v(\mathbf{x}) = v(x_1 \frac{1}{x_1}\mathbf{x}) = v(1, \frac{x_2}{x_1}, \; \ldots \; , \frac{x_n}{x_1}) + L(x_1) \; .$$

If we take (4) into consideration and write

$$f(t_2,...,t_n) = v(1,t_2,...,t_n) + u(1) \; ,$$

we get

$$u(\mathbf{x}) = u(x_1,x_2,...,x_n)$$

$$= f(\frac{x_2}{x_1}, \; \ldots \; , \frac{x_n}{x_1}) + L(x_1) \; . \tag{7}$$

Conversely, *every* function of this form, with arbitrary L satisfying (6) and with arbitrary f, satisfies (3):

$$u(r\mathbf{x}) = f(\frac{x_2}{x_1}, \; \ldots \; , \frac{x_n}{x_1}) + L(x_1) + L(r)$$

$$= u(\mathbf{x}) + L(r) \; .$$

So in this case

$$R(r) = 1, \quad P(r) = L(r) \quad (r \in \mathbb{R}_{++}) .$$ (8)

In the remaining case, where $R(r) \not\equiv 1$, we again put $\mathbf{x} = s$ into (3) and get

$$v(rs) = R(r)v(s) + v(r) \quad (r \in \mathbb{R}_{++}) .$$

As in section 2, see (2.12) and (2.13), we obtain

$$v(r) = a[M(r)-1], \quad R(r) = M(r) ,$$ (9)

where M is a multiplicative function, that is, a solution of

$$M(rs) = M(r)M(s) \quad (r,s \in \mathbb{R}_{++}) .$$ (10)

(That is why we have denoted it by M.) In view of (9) and with the notation

$$w(\mathbf{x}) = v(\mathbf{x}) + a ,$$ (11)

equation (3) goes over into

$$w(r\mathbf{x}) = M(r)w(\mathbf{x}) \quad (r \in \mathbb{R}_{++}, \ \mathbf{x} \in \mathbb{R}_{++}^n) .$$ (12)

Similarly as above, this yields w for instance in the form

$$w(x_1, x_2, ..., x_n) = w(\mathbf{x}) = w(x_1 \frac{1}{x_1}\mathbf{x})$$

$$= M(x_1)w(1, \frac{x_2}{x_1}, \ldots, \frac{x_n}{x_1})$$

$$= M(x_1)f(\frac{x_2}{x_1}, \ldots, \frac{x_n}{x_1}) .$$ (13)

where this time $f(t_2, ..., t_n) = w(1, t_2, ..., t_n)$. With (11) and (4) we have

$$u(\mathbf{x}) = u(x_1, x_2, ..., x_n)$$

$$= M(x_1)f\left(\frac{x_2}{x_1}, \ldots, \frac{x_n}{x_1}\right) + b \quad . \tag{14}$$

Conversely, every u of the form (14), with arbitrary M satisfying (10), arbitrary function f and arbitrary constant b, satisfies (2):

$$u(r\mathbf{x}) = M(r)M(x_1)f\left(\frac{x_2}{x_1}, \ldots, \frac{x_n}{x_1}\right) + b$$

$$= M(r)u(\mathbf{x}) + b[1 - M(r)] \quad .$$

So here

$$R(r) = M(r), \quad P(r) = b[1 - M(r)] \quad . \tag{15}$$

The case $v(r) \equiv 0$ (for scalar $r = (r, ..., r)$), which we have excluded from the proof of Theorem 2.1, just reduces (3) to $v(r\mathbf{x}) = R(r)v(\mathbf{x})$, similar to (12) and repeated application of this equation, that is, $R(rs)v(\mathbf{x}) = v(rs\mathbf{x}) = R(r)R(s)v(\mathbf{x})$, shows that R is multiplicative here too, *if v is not identically* 0. This gives the same solutions (14), (15) as before, only here $f(1, ..., 1) = 0$. If $v(\mathbf{x}) \equiv 0$, however, then u is constant,

$$u(\mathbf{x}) = b \quad (\mathbf{x} \in \mathbb{R}^n_{++}) \tag{16}$$

so that (2) reduces to $b = R(r)b + P(r)$ or

$$R(r) \text{ arbitrary}, \quad P(r) = b[1 - R(r)] \tag{17}$$

(where R is not necessarily multiplicative anymore). We have proved the following.

Theorem 1. *The general solutions of (2) are given by (7)-(8), (14)-(15) and (16)-(17). So the general form of laws where all independent variables have the same ratio scale, while the dependent variable has an interval scale are given by (7) and (14). Here $L : \mathbb{R}_{++} \to \mathbb{R}$ and $M : \mathbb{R}_{++} \to \mathbb{R}_{++}$ are arbitrary solutions of (6) or (10), respectively, $f : \mathbb{R}_{++}^{n-1} \to \mathbb{R}$ an arbitrary function and b an arbitrary real constant.*

(The 'law' (16) is the particular case $f = 0$ of (14) so we omitted it from the second statement.)

The case where the dependent variable also has a ratio scale, that is, it transforms by a (homogeneous) linear transformation too (while the independent variables still undergo equal linear transformations), gives the functional equation

$$u(r\mathbf{x}) = R(r)u(\mathbf{x}) \quad (r \in \mathbb{R}_{++}, \mathbf{x} \in \mathbb{R}_{++}^n;$$

$$u : \mathbb{R}_{++}^n \to \mathbb{R}_{++}, \; R : \mathbb{R}_{++} \to \mathbb{R}_{++}) \qquad (18)$$

(cf. (12)). This is simply the special case $P(r) = 0$ $(r \in \mathbb{R}_{++})$, $u(\mathbf{x}) > 0$ $(\mathbf{x} \in \mathbb{R}_{++}^n)$ of (2). So from (8) $L(r) \equiv 0$ in (7), from (15) $b = 0$ (since that was the $M(r) \not\equiv 1$ case) in (14) and from (17) $R(r) \equiv 1$ (since $b = 0$ in (16) would contradict $u(\mathbf{x}) > 0$) and f is positive valued in (7) and (14) and $b > 0$ in (16). So we have the following.

Corollary 2. *The general form of laws where all independent variables have the same ratio scale and the dependent variable also has a ratio case, that is, the general solution of (18) is given by*

$$u(\mathbf{x}) = u(x_1, x_2, ..., x_n)$$

$$= M(x_1)f(\frac{x_2}{x_1}, \ldots, \frac{x_n}{x_1}) \quad with \quad R(r) = M(r), \quad (19)$$

where $M : \mathbb{R}_{++} \rightarrow \mathbb{R}_{++}$ is an arbitrary multiplicative function and $f : \mathbb{R}_{++}^{n-1} \rightarrow \mathbb{R}_{++}$ is an arbitrary function.

(The laws (7) with $L \equiv 0$ and (16) with $b > 0$ are contained in (19) with $M(x_1) \equiv 1$ in both cases and $f = b > 0$ constant in the latter case.)

Of course these laws, containing arbitrary multiplicative and logarithmic functions, are too general. But using, say, boundedness on a proper interval from Corollary 1.5, just as we have used continuity in the proof of Theorem 2.1 we can determine all solutions u *bounded on an open n-dimensional interval I*, which certainly is a reasonable supposition for the applications we have in mind (boundedness on a set of positive measure would be sufficient but does not sound very practical). In some cases it will be enough to suppose that u be bounded *from above* on a proper interval.

For (7) and for (19) boundedness from above of u on I implies (with constant x_1) boundedness from above of f on

$$I/I_1 = \{(\frac{x_2}{x_1}, \ldots, \frac{x_n}{x_1}) \mid (x_1, x_2, ..., x_n) \in I\} \quad (20)$$

and of L and M (keeping $\mathbf{x}$ constant in (18)) on

$$I_1 = \{x_1 \mid \exists x_2, ..., x_n : (x_1, x_2, ..., x_n) \in I\} .$$

For (14) we need that u be *bounded (both from above and from below)* on the open interval I and that $f \neq 0$ on I/I_1. Indeed, then we have (again with constant x_1 in (14)) that f is bounded and different from zero on I/I_1. So by (13) there exists an $\mathbf{x}_0$ such that $w(\mathbf{x}_0) \neq 0$ and, since with u also w is bounded on I, by (12) so is M on I_1.

Now, if L and M are bounded (from above), then the method of solution applied to (2.8) and (2.13) gives, in view of Corollary 1.5,

$$L(r) = c \, \log r \qquad (21)$$

and

$$M(r) = r^c , \qquad (22)$$

respectively. [Note that boundedness from above of u and thus w may, by (12), imply boundedness *from below* of M if $w(\mathbf{x}) < 0$ (which can happen if f is negative). But this does not guarantee (22) since, as we have seen in the proof of Theorem 2.1, *every* multiplicative function is bounded from below by 0. To put it in another way, if M (or α in section 2), is bounded from below by 0 then $\phi = \log M$ need not be bounded from below anymore and we could not apply Corollary 1.5. Actually $M(r) = e^{\phi(\log r)}$ with arbitrary noncontinuous additive ϕ is everywhere positive, but of course not of the form (22). (A function is *additive* if it satisfies the Cauchy equation (1.23).) On the other hand, boundedness from above of M on an interval does carry over to ϕ.] So we have the following.

Corollary 3. *If u is supposed to be bounded on an open n-dimensional interval I in Theorem 1 and Corollary 2, then in the solutions (7), (14) and (19), f is bounded on I/I_1, as defined in (20), and in (14) either $f = 0$ on I/I_1 or M is of the form (22), while in (7) and (19) L and M are of the forms (21) and (22), respectively, where c is an arbitrary constant. In (7) and (19) it is enough to suppose that u be bounded from above on I. Then f is bounded from above on I/I_1 and L and M are still of the forms (21) and (22),* *respectively.*

We want to mention here that, while the above results give the general solutions, these solutions may also be written in different forms. We take as an example the functions u satisfying $u(r\mathbf{x}) = R(r)u(\mathbf{x})$, cf. (18). Such functions are also called *generalized homogeneous functions*. The domains need not be the same as in (18). For instance, we can consider generalized homogeneous functions $u : \mathbb{R}^n \rightarrow \mathbb{R}$ satisfying

$$u(r\mathbf{x}) = R(r)u(\mathbf{x})$$

$$\text{for all } \mathbf{x} = (x_1,...,x_n) \in \mathbb{R}^n, r \in \mathbb{R}_{++} . \tag{23}$$

As we have seen, *if* u *is not identically* 0, *then* $u[(rs)\mathbf{x}] = u[r(s\mathbf{x})]$ yields that R *is multiplicative:*

$$R(r) = M(r) , \quad M(rs) = M(r)M(s) . \tag{24}$$

Let us now substitute $r = 1/\| \mathbf{x} \| = (x_1^2 + ... + x_n^2)^{-1/2}$ into

$$u(r\mathbf{x}) = M(r)u(\mathbf{x})$$

$$(r \in \mathbb{R}_{++}, \mathbf{x} \in \mathbb{R}^n) . \tag{25}$$

We get

$$u(\mathbf{x}) = M(\| \mathbf{x} \|)u\left(\frac{\mathbf{x}}{\| \mathbf{x} \|}\right),$$

$$(\mathbf{x} \in \mathbb{R}^n, \mathbf{x} \neq 0)$$

(division of a vector by a scalar is also done componentwise), because (24) implies $M(1) = 1$ and $M(1/r) = 1/M(r)$ except if $M \equiv 0$, which, by (25), would give $u \equiv 0$. *Conversely, every function of the form*

$$u(\mathbf{x}) = M(\|\mathbf{x}\|)g\left(\frac{\mathbf{x}}{\|\mathbf{x}\|}\right)$$

$$(\mathbf{x} \in \mathbb{R}^n \setminus \{0\}), \tag{26}$$

with arbitrary g satisfies (24).

This argument does not work for $\mathbf{x} = 0$. But then (25) goes over into

$$u(0) = M(r)u(0),$$

so $u(0) = 0$ except if $M(r) \equiv 1$ in which case $u(0)$ is arbitrary. So we have proved the following.

Proposition 4. *The general solutions of (23) are given by $u(\mathbf{x}) \equiv 0$, R arbitrary and, if $u \not\equiv 0$, then by (24) and (26) with $u(0) = 0$ if $M(r) \not\equiv 1$ and $u(0) = b$ arbitrary if $M(r) \equiv 1$. Here g is an arbitrary real valued function on the n-dimensional unit sphere and M is an arbitrary real valued multiplicative function on $\mathbb{R}_{++}$.*

This gives all generalized homogeneous functions in the above sense.

When we spoke about the n-dimensional unit sphere, we meant of course

$$S = \{\mathbf{y} \mid \|\mathbf{y}\| = 1, \mathbf{y} \in \mathbb{R}^n\}$$

and this is indeed the domain where g in (26) needs to be defined. *If we want u again to be bounded on an n-dimensional open interval I, then g is bounded on*

$$I(S) = \{\frac{\mathbf{x}}{\|\mathbf{x}\|} \mid \mathbf{x} \in I \setminus \{0\}\} \tag{27}$$

and either g is 0 on $I(S)$ or

$$M(r) = r^c , \tag{28}$$

(c ᵣ real constant, as before). The problem gets intriguing if we want u to be *everywhere continuous*. If $u \not\equiv 0$, on the one hand *both g and M are continuous* and thus *M is of the form (28)*. On the other hand, we want u to be continuous also at 0. So we need

$$\lim_{\mathbf{x} \to 0} [\| \mathbf{x} \|^c g(\frac{\mathbf{x}}{\| \mathbf{x} \|})] = 0 \quad \text{if} \ c \neq 0 . \tag{29}$$

We claim that this is *true for every continuous g if $c > 0$*, while *it implies $g \equiv 0$ for $c < 0$*. Indeed, since g is continuous and is defined on the n-dimensional unit sphere, which is bounded and connected, so g is *bounded* and thus (29) is satisfied for $c > 0$. For $c < 0$, if there existed a unit vector $\mathbf{e}$ for which $g(\mathbf{e}) \neq 0$ (remember that g is — or needs to be — defined *only for unit vectors*), take $\mathbf{x} = s\mathbf{e}$. Then (29) means for $c < 0$

$$0 = \lim_{s \to 0+} s^c g(\mathbf{e}) = \pm \infty$$

(according to whether $g(\mathbf{e}) > 0$ or $g(\mathbf{e}) < 0$), which is a contradiction and proves $g \equiv 0$. Finally, if $c = 0$, we had $u(0) = b$ arbitrary, so g will be continuous at 0 only if

$$\lim_{\mathbf{x} \to 0} g(\frac{\mathbf{x}}{\| \mathbf{x} \|}) = b . \tag{30}$$

But then $g \equiv b$. Indeed, let $\mathbf{e}$ be an arbitrary unit vector and $\mathbf{x} = s\mathbf{e}$. Then (30) means $b = \lim_{s \to 0} g(\mathbf{e}) = g(\mathbf{e})$, as asserted.

So *the only everywhere continuous not identically 0 solutions of (23) (the general continuous generalized homogeneous functions) are given by*

$$u(\mathbf{x}) = \| \mathbf{x} \|^c g(\frac{\mathbf{x}}{\| \mathbf{x} \|}) \ \ for \ \ \mathbf{x} \in \mathbb{R}^n \setminus \{0\} ,$$

$$u(0) = 0 \ \ with \ \ R(r) = r^c \ (r \in \mathbb{R}_{++}), \ c > 0$$

and by

$$u(\mathbf{x}) = b \ (\mathbf{x} \in \mathbb{R}^n) \ \ with \ \ R(r) = 1 \ (r \in \mathbb{R}_{++}) \ .$$

As an example in economics, we mention the "compromise inequality measure" of Bossert-Pfingsten [86] (cf. Kolm [76]), for which they require

$$I_\mu(\mathbf{x}+\tau\,[\mu\mathbf{x}+(1-\mu)]) = I_\mu(\mathbf{x}) \tag{31}$$

($\mu \in [0,1]$ fixed; $\tau \in \mathbb{R}_+$ and $\mathbf{x} \in \mathbb{R}_+^n$ variable). For $\mu \in]0,1]$ this is transformed, with $r = 1+\tau\mu \geq 1$, $s = (1-\mu)/\mu \geq 0$, into

$$I_\mu(r(\mathbf{x}+s)-s) = I_\mu(\mathbf{x})$$

($\mathbf{x} \in \mathbb{R}_+^n$ and $r \geq 1$ variable; $s \geq 0$ fixed) . $\tag{32}$

Here *addition and subtraction of scalars to and from vectors is done componentwise*, just as is the multiplication by scalars. For example

$$r(\mathbf{x}+s)-s = r\,[(x_1,...,x_n)+s]-s$$

$$= (r(x_1+s)-s,...,r(x_n+s)-s) \ .$$

For the function u defined by

$$u(\mathbf{y}) = I_\mu(\mathbf{y}-s) \tag{33}$$

for $\mathbf{y} \geq s$ (that is, $y_i \geq s$, $i=1,...,n$), equation (32) is transformed (with $\mathbf{y} = \mathbf{x}+s$) into

$$u(r\mathbf{y}) = u(\mathbf{y}) \quad (\mathbf{y} \geq s) \ . \tag{34}$$

The additional restriction $r \geq 1$ can be lifted (as long as $r \in \mathbb{R}_{++}$) since (34) can be written as

$$u(\mathbf{z}) = u(\frac{1}{r}\ \mathbf{z}) \quad (r \geq 1, \mathbf{z} \geq s) \ ,$$

so (34) holds also for $r \leq 1$. So we have

$$u(r\mathbf{y}) = u(\mathbf{y}) \quad (\mathbf{y} \geq s, r \in \mathbb{R}_{++}) \ , \tag{35}$$

which is the particular case $R(r) = 1$ of (18) or (25), again on another domain. But, substituting $r = 1/\|\ \mathbf{y}\ \|$, we have again

$$u(\mathbf{y}) = g(\frac{\mathbf{y}}{\|\ \mathbf{y}\ \|}) \text{ for } \mathbf{y} \neq 0 \text{ and } u(0) = b \text{ (arbitrary)} \ ,$$

which indeed satisfies (35) with arbitrary g and b. In view of (33),

$$I_\mu(\mathbf{x}) = g(\frac{\mathbf{x}+s}{\|\ \mathbf{x}+s\ \|}) = g(\frac{\mu\mathbf{x}+(1-u)}{\|\ \mu\mathbf{x}+(1-\mu)\ \|})$$

$$for \mathbf{x} \neq -s = -\frac{1-\mu}{\mu} \text{ and } I_\mu(-\frac{1-\mu}{\mu}) = b$$

is the general solution of (31). (The value $-\dfrac{1-\mu}{\mu}$ is only then in the domain if $\mu = 1$). For solutions which are continuous at all $x \neq -(1-\mu)/\mu$, g is continuous (the point $-\dfrac{1-\mu}{\mu} = (\dfrac{\mu-1}{\mu}, \ \ldots \ , \dfrac{\mu-1}{\mu})$ is not very interesting for this application).

The solution is similar in the case $\mu = 0$ of (31), that is,

$$I_o(\mathbf{x}) = I_o(\mathbf{x}+\tau) \quad (\mathbf{x} \in \mathbb{R}_+^n, \tau \in \mathbb{R}_+) \ . \tag{36}$$

With $\mathbf{y} = \mathbf{x}+\tau$ this becomes

$$I_o(\mathbf{y}) = I_o(\mathbf{y}-\tau) \quad (\mathbf{y} \geq \tau \geq 0) \quad .$$

By putting into this equation for instance

$$\tau = \min(y_1,...,y_n) = m(\mathbf{y}) \quad (\mathbf{y} = (y_1,...,y_n)) \quad ,$$

we get $I_o(\mathbf{y}) = I_o(\mathbf{y}-m)$ and conversely,

$$I_o(\mathbf{x}) = I_o(x_1,...,x_n) = f(\mathbf{x}-\min(x_1,...,x_n))$$

$$(\mathbf{x} \in \mathbb{R}^n_+) \tag{37}$$

satisfies (36) with arbitrary f. So *the general solution of (36) is given by (37), where $f : \mathbb{R}^n_+ \longrightarrow \mathbb{R}$ is arbitrary. In continuous solutions f is continuous.*

Actually, (31) with $\mu \in \,]0,1[$ has been introduced as a 'compromise' between the $\mu = 0$ case (36) ('socialist' measure: inequality of income is unchanged when all incomes change by the same *amount*) and the $\mu = 1$ case of (31) ('conservative' measure: income inequality unchanged when all incomes change by the same *factor*).

The question (cf. Eichhorn-Gehrig [82]) whether there exist at all and, in particular, whether there exist continuous inequality measures which are at the same time 'socialist' and 'conservative' (say, 'liberal'), that is, satisfy both

$$I(\mathbf{x}+\tau) = I(\mathbf{x}) \quad (\mathbf{x} \in \mathbb{R}^n_+, \tau \in \mathbb{R}_+) \tag{38}$$

and

$$I(r\mathbf{x}) = I(\mathbf{x}) \quad (\mathbf{x} \in \mathbb{R}^n_+, r \in \mathbb{R}_{++}) \tag{39}$$

can be answered similarly as the problems of solving (36), (34), and (23). Write (38), just as (36), in the form

$$I(\mathbf{y}) = I(\mathbf{y}-\tau) \quad (\mathbf{y} \geq \tau \geq 0)$$

and put $\tau = \min(y_1,...,y_n) = m(\mathbf{y})$ in order to get

$$I(\mathbf{y}) = I(\mathbf{y}-m(\mathbf{y})) \quad . \tag{40}$$

Now apply (39) with $r = 1 / \| \mathbf{y}-m(\mathbf{y}) \|$ and obtain

$$I(\mathbf{y}) = I(\mathbf{y}-m(\mathbf{y}))$$

$$= I(\frac{1}{\| \mathbf{y}-m(\mathbf{y}) \|} [\mathbf{y}-m(\mathbf{y})])$$

except if

$\| \mathbf{y}-m(\mathbf{y}) \| = 0$, that is, $y_1 = ... = y_n = \min(y_1,...,y_m) = y$,

when (40) gives

$$I(\mathbf{y}) = I(y) = I(0) = b \quad \text{(arbitrary)}.$$

Conversely, one sees immediately that

$$I(\mathbf{x}) = g(\frac{1}{\| \mathbf{x}-m(\mathbf{x}) \|} [\mathbf{x}-m(\mathbf{x})])$$

$$\text{for } \mathbf{x} \neq m(\mathbf{x}) \tag{41}$$

and

$$I(\mathbf{x}) = I(x) = b \quad \text{for } \mathbf{x} = m(\mathbf{x}) = x \tag{42}$$

satisfy both (38) and (39) with arbitrary constant b and arbitrary $g:S' \to \mathbb{R}$, where, since in (41) $\min(\mathbf{x}-m(\mathbf{x})) = 0$,

$$S' = \{\mathbf{z} = (z_1,...,z_n) \mid \mathbf{z} \in \mathbb{R}_+^n, \| \mathbf{z} \| = 1, \min(z_1,...,z_n) = 0\} \quad .$$

So *there are plenty of inequality measures which are both 'conservative' and 'socialist'. But there are no noncon-stant, everywhere continuous ones.* Indeed if I *is continuous everywhere then g should be continuous, but also*

$$\lim_{\mathbf{z} \to 0} g(\frac{1}{\| \mathbf{z} \|} \mathbf{z}) = b \quad ,$$

cf. (30), so $g = b$ (constant). So we have proved the

following.

Proposition 5. *The general solutions of (38) and (39) are given by (41) and (42), where* $g : S' \rightarrow \mathbb{R}$ *is an arbitrary function,* b *is an arbitrary constant and* $m(\mathbf{x}) = m(x_1,...,x_n) = \min(x_1,...,x_n)$. *Every everywhere continuous solution of (38) and (39) is constant.*

The method used here for determining general and continuous generalized homogeneous functions and inequality measures can be used to determine the general and general continuous forms of laws where all independent variables have *the same interval scale* and the dependent variable has a ratio or interval scale. (Equation (36) and the combination of (38) and (39) are particular cases of such laws.) For details see Aczél-Roberts-Rosenbaum [86].

4. General forms of 'laws of sciences' with (partially) independent ratio and interval scales. Multiplace multiplicative and logarithmic functions. Price levels. Endomorphisms of a real field. The exponential equation. Characters.

In the previous section (cf. (3.2)) we have dealt with the case where the *same* linear (or affine) transformations were applied to each (independent) variable. In this section we look at cases where linear or affine transformations may be applied (at least partially) independently to the different independent variables.

First we will explore laws where the independent variables have *independent ratio scales* and the dependent variable has an interval scale, that is

$$u(r_1 x_1,...,r_n x_n) = R(r_1,...,r_n)u(x_1,...,x_n)$$
$$+ P(r_1,...,r_n)$$

or in vector notation

$$u(\mathbf{rx}) = R(\mathbf{r})u(\mathbf{x}) + P(\mathbf{r})$$

$$(\mathbf{r},\mathbf{x} \in I\!R^n_{++}; \ u,P:I\!R^n_{++} \rightarrow I\!R,$$

$$R:I\!R^n_{++} \rightarrow I\!R_{++}), \tag{1}$$

where $\mathbf{rx} = (r_1,...,r_n)(x_1,...,x_n) = (r_1 x_1,...,r_n x_n)$.

Both the method of solution and the solutions are exact analogues of those for (2.5) (where $n = 1$), and we have the following (cf. also the proof of Theorem 3.1).

Theorem 1. *The general solutions of (1) are given by*

$$u(\mathbf{x}) = u(x_1,...,x_n) = L(\mathbf{x}) + b = \sum_{i=1}^{n} L_i(x_i) + b$$

$$(\mathbf{x} = (x_1,...,x_n) \in \mathbb{R}^n_{++}) \tag{2}$$

with

$$R(\mathbf{r}) = 1, \quad P(\mathbf{r}) = L(\mathbf{r}) \quad (\mathbf{r} \in \mathbb{R}^n_{++}) , \tag{3}$$

by

$$u(\mathbf{x}) = u(x_1,...,x_n) = aM(\mathbf{x}) + b$$

$$= a \prod_{i=1}^{n} M_i(x_i) + b$$

$$(\mathbf{x} = (x_1,...,x_n) \in \mathbb{R}^n_{++}) \tag{4}$$

with

$$R(\mathbf{r}) = M(\mathbf{r}), \quad P(\mathbf{r}) = b[1-M(\mathbf{r})] , \tag{5}$$

and by

$$u(\mathbf{x}) = b \quad with \ arbitrary \ R$$

$$and \quad P(\mathbf{r}) = b[1-R(\mathbf{r})] . \tag{6}$$

*So the general forms of laws, where the independent vari-
ables have independent ratio scales and the dependent variable has an interval scale, are given by (2) and (4). Here
$L:\mathbb{R}^n_{++} \to \mathbb{R}$, $L_i:\mathbb{R}_{++} \to \mathbb{R}$ and $M:\mathbb{R}^n_{++} \to \mathbb{R}_{++}$,
$M_i:\mathbb{R}_{++} \to \mathbb{R}_{++}$ $(i=1,...,n)$ are arbitrary logarithmic or
multiplicative functions, respectively, that is,*

$$L(\mathbf{xy}) = L(\mathbf{x}) + L(\mathbf{y}) \ ,$$

$$M(\mathbf{xy}) = M(\mathbf{x})M(\mathbf{y}) \ \ (\mathbf{x},\mathbf{y} \in \mathbb{R}^n_{++}) \ ,$$

$$L_i(x_i y_i) = L_i(x_i) + L_i(y_i) \ ,$$

$$M_i(x_i y_i) = M_i(x_i)M_i(y_i) \ \ (x_i,y_i \in \mathbb{R}_{++}; \ i=1,...,n)$$

and a,b are arbitrary real constants.

The decompositions

$$L(\mathbf{x}) = L(x_1,...,x_n) = \sum_{i=1}^{n} L_i(x_i)$$

$$\text{and} \ \ M(\mathbf{x}) = M(x_1,...,x_n) = \prod_{i=1}^{n} M_i(x_i)$$

of the multiplace logarithmic or multiplicative functions L and M in (2) and (4) follow from Proposition 1.3. Here $(N_i,+) = (\mathbb{R}_{++},\cdot)$ and $(S,+) = (\mathbb{R},+)$ or $(S,+) = (\mathbb{R}_{++},\cdot)$, respectively (on the right hand sides we indicated here the addition and multiplication of real numbers; in Proposition 1.3, $+$ was the sign for a general operation, not necessarily addition). We did not enumerate (6) among the 'laws' u, because $u(\mathbf{x}) = b$ is a particular case both of (2) and of (4).

The case where the dependent variable also has a ratio scale is *the classical situation of dimensional analysis,* described by the functional equation (3.1) or, in vector notation,

$$u(\mathbf{rx}) = R(\mathbf{r})u(\mathbf{x})$$

$$(\mathbf{r},\mathbf{x} \in \mathbb{R}^n_{++}; \ u,R:\mathbb{R}^n_{++} \rightarrow \mathbb{R}_{++}) \ . \tag{7}$$

Again, this is the special case $P(\mathbf{r}) \equiv 0$, $u(\mathbf{x}) > 0$ of (1). So, from (3), $L(\mathbf{r}) \equiv 0$ in (2), reducing it to a constant $u(\mathbf{x}) = b > 0$, what we obtain also from (6). Finally, from (5), $b = 0$ and (4) becomes $u(\mathbf{x}) = aM(\mathbf{x})$ ($a > 0$ because

$M(\mathbf{x}) > 0$), since $M(r) \equiv 1$ would again give the constant $u(\mathbf{x}) = a + b$. So we have proved the following.

Corollary 2. *The general form of laws where all independent variables have independent ratio scales and the dependent variable also has a ratio scale, that is, the general solution of (7) is given by*

$$u(\mathbf{x}) = u(x_1,...,x_n) = aM(\mathbf{x}) = a \prod_{i=1}^{n} M_i(x_i)$$

$$\text{with} \quad R(\mathbf{r}) = M(\mathbf{r}) \quad (\mathbf{x},\mathbf{r} \in I\!R_{++}^n) , \tag{8}$$

where $M: I\!R_{++}^n \rightarrow I\!R_{++}$ *and* $M_1,...,M_n: I\!R_{++} \rightarrow I\!R_{++}$ *are arbitrary multiplicative functions and* a *is an arbitrary positive constant.*

These solutions are again too general. For (8), boundedness from above of u on a nondegenerate n-dimensional interval is sufficient to have the same property for M (since $a > 0$) and boundedness from above on one-dimensional proper intervals for M_i (by keeping x_j constant for $j \neq i, i = 1,...,n$). Then we get, by applying Corollary 1.5 as in the proof of Theorem 2.1, $M_i(x) = x^{c_i}$ $(i=1,...,n)$ and we obtain the following.

Corollary 3. *If* u *is bounded from above on a proper* n-*dimensional interval (or continuous at a point, etc.), then the general solution of (7) is given by*

$$u(\mathbf{x}) = u(x_1,...,x_n) = a \prod_{i=1}^{n} x_i^{c_i}$$

$$(\mathbf{x} = (x_1,...,x_n) \in I\!R_{++}^n) \tag{9}$$

with

$$R(\mathbf{r}) = R(r_1, \ldots, r_n) = \prod_{i=1}^{n} r_i^{c_i}$$

$$(\mathbf{r} = (r_1,...,r_n) \in \mathbb{R}^n_{++}),$$

where $a > 0$ but otherwise $a, c_1, ..., c_n$ are arbitrary constants. So (9) gives the general laws, continuous at a point or bounded from above on a proper n-dimensional interval, where the independent variables have independent ratio scales and the dependent variable has also a ratio scale.

This is *the classical result of dimensional analysis* see, for instance Ellis [66] and Luce [64], where differentiability or continuity was supposed, respectively, and Mohr [51] where one of the conditions, each of which guarantees that (9) is the general solution of (7), is given as boundedness from below on a proper interval. As we have seen in the proof of Corollary 3.3, this does not guarantee

$$M_i(x_i) = x_i^{c_i} \quad (i=1,...,n),$$

$$M(\mathbf{x}) = \prod_{i=1}^{n} x_i^{c_i} . \tag{10}$$

But the alternative condition in Mohr [51], boundedness from above of u on a proper interval, is sufficient for (9) to be the general solution of (7). However, for the same reason as above, not even boundedness from above of u is sufficient to establish (10) in (4). Indeed, there a may be negative and, if $\lambda < b$ for the upper bound λ of u, then we have at most $M(\mathbf{x}) > 0$ which, as we have seen, does not imply (10). However, if u is supposed to be bounded (from both sides) on a proper interval (or continuous at a point, etc.), then we get (10). So the following holds.

Corollary 4. *If* u *is supposed to be bounded on an* n-*dimensional proper interval (or continuous at a point, etc.) then in Theorem 1*

$$u(x_1,...,x_n) = \sum_{i=1}^{n} c_i \log x_i + b ,$$

or

$$u(x_1,...,x_n) = a \prod_{i=1}^{n} x_i^{c_i} + b ,$$

where $a,b,c_1,...,c_n$ *are arbitrary real constants.*

This too has been obtained by Luce [64] under the supposition that u is everywhere continuous.

Our next case is where the independent variables all have *interval scales with the same unit but with independent zeros*, while the dependent variable also has an interval scale. So we have to solve the functional equation

$$u(rx_1+p_1,...,rx_n+p_n) = R(r,p_1,...,p_n)u(x_1,...,x_n)$$
$$+ P(r,p_1,...,p_n)$$

or, in vector notation,

$$u(r\mathbf{x}+\mathbf{p}) = R(r,\mathbf{p})u(\mathbf{x}) + P(r,\mathbf{p})$$

$$(\mathbf{x},\mathbf{p} \in \mathbb{R}^n, r \in \mathbb{R}_{++}; u : \mathbb{R}^n \to \mathbb{R},$$

$$R : \mathbb{R}_{++} \times \mathbb{R}^n \to \mathbb{R}_{++}, P : \mathbb{R}_{++} \times \mathbb{R}^n \to \mathbb{R}) \quad (11)$$

In order to solve this equation, we substitute this time $\mathbf{x} = 0$ and subtract the equation thus obtained from (11) which gives

$$w(r\mathbf{x}+\mathbf{p}) = R(r,\mathbf{p})w(\mathbf{x}) + w(\mathbf{p}) \tag{12}$$

for

$$w(\mathbf{x}) = u(\mathbf{x}) - u(0) \ . \tag{13}$$

Repeated use of (12) yields

$$R(rs,r\mathbf{q}+\mathbf{p})w(\mathbf{x}) + w(r\mathbf{q}+\mathbf{p}) = w[rs\mathbf{x}+(r\mathbf{q}+\mathbf{p})]$$

$$= w[r(s\mathbf{x}+\mathbf{q})+\mathbf{p}]$$

$$= R(r,\mathbf{p})R(s,\mathbf{q})w(\mathbf{x}) + R(r,\mathbf{p})w(\mathbf{q}) + w(\mathbf{p})$$

that is,

$$R(rs,r\mathbf{q}+\mathbf{p}) = R(r,\mathbf{p})R(s,\mathbf{q})$$

$$(r,s \in \mathbb{R}_{++}; \ \mathbf{p},\mathbf{q} \in \mathbb{R}^n) \tag{14}$$

except if w and so, by (13), also u *is constant.* Defining new functions M and E by $M(r) = R(r,0)$, $E(\mathbf{p}) = R(1,\mathbf{p})$, we get from (14) with $\mathbf{p} = \mathbf{q} = 0$ that M is multiplicative

$$M(rs) = M(r)M(s) \ \ (r,s \in \mathbb{R}_{++}) \tag{15}$$

and, with $r = s = 1$, that E is 'exponential', that is, satisfies

$$E(\mathbf{p}+\mathbf{q}) = E(\mathbf{p})E(\mathbf{q}) \ \ (\mathbf{p},\mathbf{q} \in \mathbb{R}^n) \ . \tag{16}$$

Of course, with R also M and E will be *positive* valued.' Further use of (14) yields

$$R(s,\mathbf{p}) = R(1,\mathbf{p})R(s,0) = M(s)E(\mathbf{p}) \tag{17}$$

and from (17) and (14)

$$M(r)E(r\mathbf{q}) = R(r,r\mathbf{q}) = R(r,0)R(1,\mathbf{q}) = M(r)E(\mathbf{q}) \ .$$

Since $M(r) > 0$, this gives $E(r\mathbf{q}) = E(\mathbf{q})$, in particular $E(2\mathbf{q}) = E(\mathbf{q})$. But from (16), $E(2\mathbf{q}) = E(\mathbf{q})^2$. So $E(\mathbf{q}) = 1$ (since $E(\mathbf{q}) > 0$) and, by (17),

$$R(s,\mathbf{p}) = M(s) \,,$$

that is, R is *independent of* $\mathbf{p}$ and (12) becomes

$$w(r\mathbf{x}+\mathbf{p}) = M(r)w(\mathbf{x}) + w(\mathbf{p})$$

$$(\mathbf{x},\mathbf{p} \in \mathbb{R}^n, \, r \in \mathbb{R}_{++}) \ .$$

Putting here $\mathbf{p} = 0$ or $r = 1$, we get

$$w(r\mathbf{x}) = M(r)w(\mathbf{x}) \ \ (r \in \mathbb{R}_{++}, \, \mathbf{x} \in \mathbb{R}^n) \qquad (18)$$

(since $w(0) = 0$ by (13)) and

$$w(\mathbf{x}+\mathbf{p}) = w(\mathbf{x}) + w(\mathbf{p}) \ \ (\mathbf{x},\mathbf{p} \in \mathbb{R}^n) \qquad (19)$$

(since, $M(1) = 1$ by (15) in view of $M(r) > 0$), respectively. Therefore

$$M(r+s)w(\mathbf{x}) = w[(r+s)\mathbf{x}] = w(r\mathbf{x}+s\mathbf{x})$$

$$= w(r\mathbf{x}) + w(s\mathbf{x})$$

$$= [M(r)+M(s)]w(x) \ .$$

So, if $w \not\equiv 0$ (u not constant), then

$$M(r+s) = M(r) + M(s) \ \ (r,s > 0) \ . \qquad (20)$$

So we have now two equations, (15) and (20) for M. We can solve this system *without any regularity conditions*. From (15) [just as from (2.13)] we get, even if we did not know it before, that

$$M(r) = M(\sqrt{r}\,\sqrt{r}) = M(\sqrt{r})^2 \geq 0 \ .$$

In view of this, (20) implies by Corollary 1.8

$$M(r) = \alpha r \ \ (r \in \mathbb{R}_{++}) \ .$$

But this satisfies (15) only if $\alpha^2 = \alpha$. So *the only common solutions of (15) and (20) are*

$$M(r) \equiv 0 \quad and \quad M(r) = r \quad (r \in \mathbb{R}_{++}) \tag{21}$$

Of course, we have $M(r) > 0$ so only the second solution is of use for us.

Thus $M(r) = r$ and (18) becomes

$$w(r\mathbf{x}) = rw(\mathbf{x}) \quad (r \in \mathbb{R}_{++}, \mathbf{x} \in \mathbb{R}^n) \tag{22}$$

for *positive* r. Since, however, (19) implies

$$w(0) = 0 \quad and \quad w(-\mathbf{x}) = -w(\mathbf{x})$$

(just as (1.26) followed from (1.23)), we have

$$w(t\mathbf{x}) = tw(\mathbf{x}) \quad \text{for all} \quad t \in \mathbb{R} \quad \text{and all} \quad \mathbf{x} \in \mathbb{R}^n \quad . \tag{23}$$

Together with (19) this means that w is *linear*, that is,

$$w(\mathbf{x}) = w(x_1,...,x_n) = w(\sum_{i=1}^{n} x_i \mathbf{e}_i)$$

$$= \sum_{i=1}^{n} w(\mathbf{e}_i) x_i = \sum_{i=1}^{n} a_i x_i \tag{24}$$

and we have from (13)

$$u(\mathbf{x}) = u(x_1,...,x_n) = \sum_{i=1}^{n} a_i x_i + b$$

$$(\mathbf{x} = (x_1,...,x_n) \in \mathbb{R}^n) \quad . \tag{25}$$

This includes also the constant solution $u = b$. Substitution of (25) into (11) gives

$$R(r,\mathbf{p}) = r, \quad P(r,\mathbf{p}) = b(1-r) + \sum_{i=1}^{n} a_i p_i$$

$$(r \in \mathbb{R}_{++}, \mathbf{p} \in \mathbb{R}^n) \tag{26}$$

if u is not constant; otherwise

$$u(\mathbf{x}) = b, \ R \text{ arbitrary}, \ P(r,\mathbf{p}) = b\,[1-R(r,\mathbf{p})]$$

$$(r \in \mathbb{R}_{++}; \ \mathbf{x},\mathbf{p} \in \mathbb{R}^n) \quad . \tag{27}$$

So we have proved the following.

Theorem 5. *The general solution (without any regularity supposition) of (11) are given by (25)-(26) and (27). So the general form of laws, where the independent variables have interval scales with the same unit but with independent zeros and where the dependent variable has an interval scale too, is given by (25), where $a_1,...,a_n$, b are arbitrary constants.*

With the same kind of scales for the independent variables, but with a *ratio* scale for the dependent variable, we have the equation

$$u(r\mathbf{x}+\mathbf{p}) = R(r,\mathbf{p})u(\mathbf{x}) \ (\mathbf{x},\mathbf{p} \in \mathbb{R}^n, \ r \in \mathbb{R}_{++};$$

$$u:\mathbb{R}^n \rightarrow \mathbb{R}_{++}, \ R:\mathbb{R}_{++} \times \mathbb{R}^n \rightarrow \mathbb{R}_{++}) \tag{28}$$

which, of course, is the case $P(r,\mathbf{p}) = 0$, $u(\mathbf{x}) > 0$ of (11). While in (26) $P(r,\mathbf{p}) = 0$ gives $a_1 =...= a_n = b = 0$, reducing (25) to $u(\mathbf{x}) = 0$ in contradiction to $u(x) > 0$, from (27) we get $R(r,\mathbf{p}) = 1$, $u(\mathbf{x}) = b > 0$. So we have proved the following.

Corollary 6. *The only solution of (28) (without any prior regularity supposition) is*

$$u(\mathbf{x}) = b > 0, \quad R(r,\mathbf{p}) = 1 \ .$$

So the only 'law', where the independent variables have interval scales with the same unit and independent zeros and where the dependent variable has a ratio scale, is a positive constant.

We have several byproducts of the proof of Theorem 5. *The pair of functional equations (19), (22) is satisfied* in economics *by the price level.* We have proved, *without any further supposition,* that *it has to be of the linear form (24).* In Eichhorn-Voeller [76] (cf. Eichhorn [78]) the price level w is defined on $I\!R^n_+$ rather than $I\!R^n$. Exactly as above we can show that *the general solution* $w: I\!R^n_+ \longrightarrow I\!R$ *of*

$$w(\mathbf{x}+\mathbf{y}) = w(\mathbf{x}) + w(\mathbf{y}) \quad (\mathbf{x},\mathbf{y} \in I\!R^n_+)$$

and

$$w(r\mathbf{x}) = rw(\mathbf{x}) \quad (r \in I\!R_{++}, \mathbf{x} \in I\!R^n_+)$$

is of the form

$$w(\mathbf{x}) = w(x_1,...,x_n) = \sum_{i=1}^{n} a_i x_i$$

$$(\mathbf{x} = (x_1,...,x_n) \in I\!R^n_+)$$

where $a_1,...,a_n$ *are arbitrary real constants.* But in Eichhorn-Voeller [76] and Eichhorn [78] also *the values of* w *are supposed to be nonnegative* (naturally: neither the individual prices nor the price level is negative). *Then* $w(r\mathbf{x}) = rw(\mathbf{x})$ *need not to be supposed at all: The general solution* $w: I\!R^n_+ \longrightarrow I\!R_+$ *of*

$$w(\mathbf{x}+\mathbf{y}) = w(\mathbf{x}) + w(\mathbf{y}) \quad (\mathbf{x},y \in I\!R^n_+)$$

is of the form

$$w(\mathbf{x}) = w(x_1,...,x_n) = \sum_{i=1}^{n} a_i x_i \quad (\mathbf{x} \in I\!R^n_+)$$

where $a_1,...,a_n$ *are arbitrary nonnegative constants.* Indeed $w: I\!R^n_+ \longrightarrow I\!R_+$, that is,

$$w(\mathbf{x}) \geq 0 \text{ for } \mathbf{x} \in \mathbb{R}^n_+$$

means that w is bounded from below on the n-dimensional proper interval $\mathbb{R}^n_+$ and the result follows from Corollary 1.9, since $\sum_{i=1}^n a_i x_i \geq 0$ for $x_1 \geq 0,...,x_n \geq 0$ implies, of course, $a_1 \geq 0,...,a_n \geq 0$. If we suppose also that $w(\mathbf{x}) > 0$ for all $\mathbf{x} \in \mathbb{R}^n_{++}$, then $a_1 > 0,...,a_n > 0$.

Another definition of price levels $u : \mathbb{R}^n_+ \to \mathbb{R}_+$, also in Eichhorn-Voeller [76] supposes that u be increasing, strictly increasing on $\mathbb{R}^n_{++}$, in each variable and the analog

$$u(\mathbf{rx}) = R(\mathbf{r})u(\mathbf{x}) \quad (\mathbf{x,r} \in \mathbb{R}^n_+)$$

of (7) be satisfied. Then it is immediate, as in the proof of Corollary 3, that

$$u(\mathbf{x}) = a \prod_{i=1}^n x_i^{c_i} \quad (\mathbf{x} \in \mathbb{R}^n_+)$$

is the general solution, this time with $a > 0$, $c_1 > 0,...,c_n > 0$. Only the last inequalities and the possibility that some x_i may be 0 need to be discussed. If we first restrict ourselves to $\mathbf{x,r} \in \mathbb{R}^n_{++}$, then we have essentially (7) (the possibility of 0 values for u does not change anything) and thus, by Corollary 2, (8) holds. Since $u(\mathbf{x}) \geq 0$ and u is strictly increasing, we have indeed (9) with $a > 0$, $c_1 > 0,...,c_n > 0$. If some $x_i = 0$ then, because u was supposed to be increasing (though not strictly increasing) also in this situation, u has to be 0 and the above form of u does not change.

The fact that the general solution of the pair of functional equations (15), (20) is given by (21) can be expressed in the following way. (Note that we did not use there the prior information that M is positive valued.)

Proposition 7. *Every homomorphism from $(\mathbb{R}_{++},+,\cdot)$ into $(\mathbb{R},+,\cdot)$ is trivial (either the zero or the identity mapping).*

A homomorphism of a structure into itself is an *endomorphism*. Using Proposition 7 or its proof we get (without even using the existence of inverses under multiplication) the following.

Corollary 8. *The real field $(\mathbb{R},+,\cdot)$ has only trivial endomorphisms.*

We come to our final pair of laws. (With this we will have taken care of 8 of the 12 possible laws with ratio or interval scales where for all independent variables we have the same types of allowable transformations. For the remaining four cases and also for logarithmic scales see Aczél-Roberts-Rosenbaum [86]). Here all *independent variables have* completely *independent interval scales* and the dependent variable will have first an interval scale and later a ratio scale. In the first case we have the equation

$$u(r_1 x_1 + p_1,...,r_n x_n + p_n) = R(r_1,...,r_n,p_1,...,p_n)u(x_1,...,x_n)$$
$$+ P(r_1,...,r_n,p_1,...,p_n) \tag{29}$$

or in vector notation

$$u(\mathbf{rx+p}) = R(\mathbf{r,p})u(\mathbf{x}) + P(\mathbf{r,p})$$
$$(\mathbf{x,p} \in \mathbb{R}^n, \mathbf{r} \in \mathbb{R}^n_{++}; u : \mathbb{R}^n \rightarrow \mathbb{R},$$
$$R : \mathbb{R}^n_{++} \times \mathbb{R}^n \rightarrow \mathbb{R}_{++}, P : \mathbb{R}^n_{++} \times \mathbb{R}_n \rightarrow \mathbb{R}) . \tag{30}$$

But *(30) implies (11)* as we see if we put $\mathbf{r} = (r,...,r) = r$ into (30). So *all solutions u of (30) have to* be of the form (25). Substitution of (25) into the left hand side of (29) (which is the detailed form of (30)) gives

$$\sum_{i=1}^{n} a_i(r_i x_i + p_i) + b = \sum_{i=1}^{n} r_i a_i x_i + \sum_{i=1}^{n} a_i p_i + b \ .$$

By comparing the coefficient of x_i $(i=1,...,n)$ with that on the right hand side

$$R(r_1,...,r_n,p_1,...,p_n)(\sum_{i=1}^{n} a_i x_i + b) + P(r_1,...,r_n,p_1,...,p_n)$$

of (29), we get

$$R(r_1,...,r_n,p_1,...,p_n)a_i = r_i a_i \ .$$

But we can have

$$R(r_1,...,r_n,p_1,...,p_n) = r_i \quad \text{only for one} \quad i(= j) \ ,$$

so

$$a_i = 0 \quad \text{for} \quad i \neq j$$

and, writing $a_j = a$, we get

$$u(\mathbf{x}) = u(x_1,...,x_n) = a x_j + b$$

$$\text{for one} \quad j \in \{1,...,n\} \ . \tag{31}$$

Substitution into (30) gives

$$u(\mathbf{rx+p}) = a(r_j x_j + p_j) + b$$

$$= r_j u(\mathbf{x}) + a p_j + b(1-r)$$

that is, if u is not constant,

$$R(r_1,...,r_n,p_1,...,p_n) = r_j,$$

$$P(r_1,...,r_n,p_1,...,p_n) = a p_j + b(1-r) \ . \tag{32}$$

If u is constant then, of course,

$$u(\mathbf{x}) = b, \ R \text{ arbitrary,}$$

$$P(\mathbf{r},\mathbf{p}) = b\left[1-R(\mathbf{r},\mathbf{p})\right] \ . \tag{33}$$

If we again have independent interval scales for the independent variables, but a ratio scale for the dependent variable, then the equation is

$$u(\mathbf{rx}+\mathbf{p}) = R(\mathbf{r},\mathbf{p})u(\mathbf{x})$$

$$(\mathbf{x},\mathbf{p} \in I\!\!R^n, \ \mathbf{r} \in I\!\!R^n_{++}; \ u:I\!\!R^n \rightarrow I\!\!R_{++},$$

$$R:I\!\!R^n_{++} \times I\!\!R^n \rightarrow I\!\!R_{++}) \ . \tag{34}$$

Here too, with $\mathbf{r} = (r,...,r) = r$, equation (34) implies (28) which had only (positive) constant solutions. So

$$u(\mathbf{x}) = b > 0, \quad R(\mathbf{r},\mathbf{p}) = 1 \ , \tag{35}$$

which, of course, satisfies (34). We finally have the following.

Proposition 9. *The general solutions of (30) are given by (31) - (32) and by (33), where a and b are arbitrary constants, while (34) has only the constant solution (35). So the general form of laws, where the independent variables have independent interval scales and the dependent variable has also an interval scale, is given by (31) while, with the same kind of scales for the independent variables, if the dependent variable has a ratio scale then the only 'law' is constant.*

This result has been obtained by Luce [64] under the supposition that u is (everywhere) continuous. — It is worth repeating that *no regularity assumptions* whatsoever were made in Theorem 5, Corollaries 6, 8 and Propositions 7 and 9 but the solutions turned out to be *very regular* (affine). — A

further remark is that the above methods of solution can also
be applied in more general situations, for instance when the
vector **x** is multiplied by a matrix rather than by a vector or
scalar (which are diagonal matrices) and also when the under-
lying structure is more general than the n-dimensional real
space and its subsets.

For the sake of completeness we give also the complete
solution of the *'exponential equation'*

$$E(\mathbf{p}+\mathbf{q}) = E(\mathbf{p})E(\mathbf{q})$$

$$(\mathbf{p}=(p_1,...,p_n)\in\mathbb{R}^n, \mathbf{q}=(q_1,...,q_n)\in\mathbb{R}^n; \ E:\mathbb{R}^n\rightarrow\mathbb{R}) \quad (16)$$

(Without the added equation $E(2\mathbf{q}) = E(\mathbf{q})$ it is *not* true that
(16) has only constant solutions.) Let $\mathbf{x} = (x_1,...,x_n) \in \mathbb{R}^n_{++}$
and $\mathbf{y} = (y_1,...,y_n) \in \mathbb{R}^n_{++}$ be arbitrary and put $p_i = \log x_i$,
$q_i = \log y_i$ and

$$M(\mathbf{x}) = M(x_1,...,x_n)$$

$$= E(\log x_1, \ldots, \log x_n) \quad (36)$$

into (16) in order to get

$$M(\mathbf{xy}) = M(\mathbf{x})M(\mathbf{y})$$

$$(\mathbf{x},\mathbf{y} \in \mathbb{R}^n_{++}; \ M:\mathbb{R}^n_{++} \rightarrow \mathbb{R}) \quad . \quad (37)$$

As we have seen, (37) implies

$$M(\mathbf{x}) = \prod_{i=1}^{n} M_i(x_i) \ (\mathbf{x} = (x_1,...,x_n) \in \mathbb{R}^n_{++}) \ ,$$

where

$$M_i(x_i y_i) = M_i(x_i)M_i(y_i) \ (x_i,y_i \in \mathbb{R}_{++}) \ .$$

(In Theorem 1 and Corollary 2, M was positive valued but we can show, as in the proof of Theorem 5 and Proposition 7, that *every solution of (37) is either positive valued or identically zero*). Furthermore, *if M is not identically 0 and bounded from above on an n-dimensional proper interval (or continuous at a point, etc.) then* (see Corollaries 3, 4 or 1.9)

$$M(\mathbf{x}) = \prod_{i=1}^{n} x_i^{c_i}$$

for some real constants $c_1,...,c_n$. In view of (36) we have the following.

Proposition 10. *The general not identically 0 solution $E: \mathbb{R}^n \to \mathbb{R}$ of (16), bounded from above on an n-dimensional proper interval (or continuous at a point, etc.) is given by*

$$E(\mathbf{p}) = e^{\sum_{i=1}^{n} c_i p_i} \quad (\mathbf{p} = (p_1,...,p_n) \in \mathbb{R}^n) \ ,$$

where $c_1,...,c_n$ are arbitrary real constants.

If, in particular, $n = 1$ or $n = 2$, we get

$$E(p) = e^{cp}$$

and

$$E(p_1,p_2) = e^{c_1 p_1 + c_2 p_2}$$

respectively. The latter result can be reformulated as follows. *The general not identically zero solution $E: \mathbb{C} \to \mathbb{R}$ ($\mathbb{C}$ is the set of complex numbers) of*

$$E(z+w) = E(z)E(w)$$

$$(z = z_1 + iz_2 \in \mathbb{C}, \ w = w_1 + iw_2 \in \mathbb{C}) \ ,$$

continuous at a point (or bounded from above on a proper

complex interval), is given by

$$E(z) = e^{c_1 z_1 + c_2 z_2} = e^{az + \overline{a}\overline{z}}$$

where c_1, c_2 are arbitrary real constants, a is an arbitrary complex constant and

$$\overline{z} = \overline{z_1 + i z_2} = z_1 - i z_2 \ .$$

The solution is a bit more difficult if *E maps $\mathbb{R}$ into the set of nonzero complex numbers:*

$$E(x+y) = E(x)E(y) \quad (x,y \in \mathbb{R}, \ E:\mathbb{R} \to \mathbb{C} \setminus \{0\}) \ . \quad (38)$$

If $\mathbb{R}$ is replaced by a topological group in (38) then E is called a *character*; so (38) describes the characters on $\mathbb{R}$. If the character E is *bounded on $\mathbb{R}$* (there exist a μ such that $|E(x)| < \mu$ for all $x \in \mathbb{R}$), then

$$|E(x)| = 1 \ . \quad (39)$$

Indeed, again *E is nowhere* 0 and, just as (1.25) followed from (1.23), we have from (38)

$$|E(nx)| = |E(x)|^n$$

for all positive and negative integer n, so, if $|E(x)| \neq 1$, then E is not bounded. Therefore

$$E(x) = e^{i\psi(x)} \quad (x \in \mathbb{R}, \ \psi: \mathbb{R} \to \mathbb{R}) \ . \quad (40)$$

We will now *suppose that E is continuous* (measurability would be enough). Integrating (38), we get

$$\int_0^b E(x+y)dy = E(x)\int_0^b E(y)dy$$

or, with $t = x+y$,

$$\int\limits_{x}^{x+b} E(t)dt = c(b)E(x) \qquad (41)$$

where $c(b) = \int\limits_{0}^{b} E(y)dy$. Since E is continuous and $E \not\equiv 0$, it

is impossible that $\int\limits_{0}^{s} E(y)dy \equiv 0$. So there exist a b such that

$c(b) = \int\limits_{0}^{b} E(y)dy \neq 0$. Thus from (41)

$$E(x) = \frac{1}{c(b)}(\int\limits_{0}^{x+b} E(t)dt - \int\limits_{0}^{x} E(t)dt) \ . \qquad (42)$$

Since E is continuous, the right hand side of (42) is differentiable so E is also differentiable. Differentiating (38) and setting then $x = 0$, $E(0) = a$, we obtain

$$E'(x+y) = E(x)E'(y), \ \ E'(x) = aE(x) \ ,$$

$$E(x) = Ae^{ax} \ .$$

Putting this back into (38) and (39) we get $A = 1$, $a = i\gamma$ and the following.

Proposition 11. *The general continuous character on $\mathbb{R}$ is given by $E(x) = e^{i\gamma x}$ (γ an arbitrary real constant).*

As mentioned before, the continuity assumption may be replaced by (Lebesgue) measurability.

5. Pexider's equation and its extension.
Quasi-extension of Cauchy's equation.
Determination of all generalized
Hicks-neutral production functions.

We will deal first with the *Pexider equation* (an equation with three unknown functions)

$$h(xy) = f(x)g(y) \quad (x,y \in N) \quad . \tag{1}$$

(of which (1.5), (4.7) and others are special cases) for $f,g,h:N \to G$ in similar generality as with (1.19) in Proposition 1.3: We will suppose that $(N,\cdot)$ is a groupoid with a neutral element and $(G,\cdot)$ a group (so, unlike the situation for (1.19), G is supposed to have neutral and inverse elements but it will not be supposed in this general set-up that G is commutative). As in (1) we will continue to omit the '·' sign; we will also denote, as usual, the inverse of $c \in G$ by c^{-1}.

Let us call e the neutral element of N and substitute in (1) $y = e$ or $x = e$. Then, with the notations $f(e) = a$, $g(e) = b$ we get

$$f(x) = h(x)b^{-1} \quad (n \in N) \tag{2}$$

and

$$g(y) = a^{-1}h(y) \quad (y \in N) \ , \tag{3}$$

respectively. Plugging these back into (1), and taking the associativity in G into consideration, we get

71

$$h(xy) = h(x)b^{-1}a^{-1}h(y) \ .$$

With the new function $\phi: N \longrightarrow G$, defined by

$$\phi(x) = a^{-1}h(x)b^{-1} \ (x \in N) \ , \tag{4}$$

we have

$$\phi(xy) = \phi(x)\phi(y) \ (x,y \in N) \ . \tag{5}$$

In view of (4), (2) and (3),

$$h(x) = a\phi(x)b, \ f(x) = a\phi(x),$$

$$g(x) = \phi(x)b \ (x \in N) \ . \tag{6}$$

Conversely, using the associativity again, (1) is satisfied by all f,g,h of the form (6), with arbitrary constants $a,b \in N$ and an arbitrary solution ϕ of (5). Triples of mappings $f,g,h: N \longrightarrow G$ satisfying (1) are often called *homotopisms* of $(N,\cdot)$ into $(G,\cdot)$ while a $\phi: N \longrightarrow G$ satisfying (5) is, of course, a homomorphism. So we have the following.

Proposition 1. *Let $(N,\cdot)$ be a groupoid with neutral element and $(G,\cdot)$ a group. The general solution $f,g,h: N \longrightarrow G$ of (1) is given by (6), where ϕ is an arbitrary solution of (5) and $a,b \in G$ are arbitrary constants. In other words, every homotopism of $(N,\cdot)$ into $(G,\cdot)$ is of the form (6) where $\phi: N \longrightarrow G$ is a homomorphism and $a,b \in N$ are constants.*

Corollary 2. *The general solutions $f,g,h: \mathbb{R} \longrightarrow \mathbb{R}$ of*

$$h(x+y) = f(x) + g(y) \ (x,y \in \mathbb{R}) \tag{7}$$

are given by

$$f(x) = \phi(x) + \alpha, \; g(x) = \phi(x) + \beta,$$

$$h(x) = \phi(x) + \alpha + \beta \;\; (x \in \mathbb{R}), \qquad (8)$$

where α, β are arbitrary real constants and ϕ is an arbitrary solution of the Cauchy equation

$$\phi(x+y) = \phi(x) + \phi(y) \;\; (x,y \in \mathbb{R}) \; . \qquad (9)$$

If f or g or h is continuous at a point or bounded from one side on a proper interval (or on a set of positive measure, etc.) then the general solution of (7) is given by

$$f(x) = cx + \alpha, \; g(x) = cx + \beta,$$

$$h(x) = cx + \alpha + \beta \;\; (x \in \mathbb{R}) \; ,$$

where α, β, c are arbitrary real constants.

Corollary 3. *The general solutions $f, g, h \colon \mathbb{R}_{++} \to \mathbb{R}^{*} = \mathbb{R} \setminus \{0\}$ (the set of nonzero reals) of*

$$h(xy) = f(x)g(y) \;\; (x,y \in \mathbb{R}_{++}) \qquad (10)$$

are given by

$$f(x) = aM(x), \; g(x) = bM(x),$$

$$h(x) = abM(x) \;\; (x \in \mathbb{R}_{++}) \; , \qquad (11)$$

where a, b are arbitrary real constants different from 0 and $M \colon \mathbb{R}_{++} \to \mathbb{R}_{++}$ is an arbitrary multiplicative function:

$$M(xy) = M(x)M(y) \;\; (x,y \in \mathbb{R}_{++}) \; . \qquad (12)$$

If at least one of f, g, h is continuous at a point or bounded on a proper interval, then the general solution of (10) is given by

$$f(x) = ax^c, \; g(x) = bx^c,$$

$$h(x) = abx^c \; (x \in \mathbb{R}_{++}) \; , \tag{13}$$

where ab $\neq 0$ but a,b,c are otherwise arbitrary real constants.

(As we have seen in sections 2 and 4, (12) implies $M(x) = M(\sqrt{x})^2 \geq 0$ and, since $M(\sqrt{x}) \neq 0$, so, if $M: \mathbb{R}_{++} \rightarrow \mathbb{R}^* = \mathbb{R} \setminus \{0\}$, then necessarily $M: \mathbb{R}_{++} \rightarrow \mathbb{R}_{++}$).

In the proof of Proposition 1 we needed only that $a = f(e)$ and $b = g(e)$ have inverses. If in (10) $f,g,h: \mathbb{R}_{++} \rightarrow \mathbb{R}$ (0 is not excluded as a function value now) but $f(1) \neq 0$, $g(1) \neq 0$ then we still get (11) as general solutions. If $f(1) = 0$ or $g(1) = 0$ then from (10) $h(x) = 0$ (for all $x \in \mathbb{R}_{++}$) and this in turn implies either $f(x) \equiv 0$ or $g(y) \equiv 0$ (and then g or f is arbitrary, respectively). So, *if* $f,g,h: \mathbb{R}_{++} \rightarrow \mathbb{R}$, *then* $g(x) = h(x) = 0 \; (x \in \mathbb{R}_{++})$, f *arbitrary and* $f(x) = h(x) = 0 \; (x \in \mathbb{R}_{++})$, g *arbitrary are added to the solutions (11) and (13)* (in the latter case appropriate regularity conditions may have to be imposed upon the "arbitrary" g or h).

We have seen in section 1 that the Cauchy equation can be extended from certain subsets (triangles, hexagons, quarter-planes) of $\mathbb{R}^2$ to $\mathbb{R}^2$. This is not so for all subsets, not even for all regions (connected open sets). Take for instance the open rectangle $]4,5[\times]6,7[$. Then the function f defined by

$$f(x) = \begin{cases} x+1 & \text{if} \quad x \in]4,5[\\ x+2 & \text{if} \quad x \in]6,7[\\ x+3 & \text{if} \quad x \in]10,12[\end{cases} \tag{14}$$

satisfies the restricted Cauchy equation $f(x+y) = f(x) + f(y)$ *on* $]4,5[\times]6,7[$ *that is, for all*

$x \in \,]4,5[$, $y \in \,]6,7[$ (the domain $]4,5[\times]6,7[$ of this equation is a region even though the domain of f in (14) is not connected). However, *there exists no solution* $\phi:\mathbb{R} \rightarrow \mathbb{R}$ *of the Cauchy equation*

$$\phi(x+y) = \phi(x) + \phi(y)$$

$$\text{for all} \quad (x,y) \in \mathbb{R}^2 \tag{15}$$

which is an extension of f. Indeed f as defined in (14) is bounded, even continuous on $]4,5[$. So, if ϕ were an extension of f then it too would be bounded on $]4,5[$, and by Corollary 1.5 we would have $\phi(x) = \alpha x$ for some α, and then $f(x) = \phi(x) = \alpha x$ on $]4,5[$, contrary to (14). However, if we define f,g,h by

$$f(x) = x + 1 \quad \text{for} \quad x \in \,]4,5[\;,$$

$$g(x) = x + 2 \quad \text{for} \quad x \in \,]6,7[$$

and

$$h(x) = x + 3 \quad \text{for} \quad x \in \,]10,12[\quad,$$

then these functions satisfy the restricted Pexider equation

$$h(x+y) = f(x) + g(y)$$

$$\text{for all} \quad (x,y) \in \,]4,5[\times]6,7[$$

and there *do exist* (unique) *extensions* $F:\mathbb{R} \rightarrow \mathbb{R}$ of f, $G:\mathbb{R} \rightarrow \mathbb{R}$ of g and $H:\mathbb{R} \rightarrow \mathbb{R}$ of H which satisfy the Pexider equation

$$H(x+y) = F(x) + G(y)$$

$$\text{for all} \quad (x,y) \in \mathbb{R}^2\;,$$

(namely $F(x) = x + 1$, $G(x) = x + 2$, $H(x) = x + 3$ for all $x \in \mathbb{R}$). This suggests that it is more natural to consider extending Pexider's equation than Cauchy's (cf. Radó-Baker [86]).

For $S \subseteq \mathbb{R}^2$ ($\mathbb{R}$ may be replaced here by other sets, say real or complex linear topological spaces and the range of f, g, h by an arbitrary abelian group) we define

$$\begin{aligned}
S_x &= \{x \mid \exists y : (x,y) \in S\}, \\
S_y &= \{y \mid \exists x : (x,y) \in S\}, \\
S_{x+y} &= \{x+y \mid (x,y) \in S\}.
\end{aligned} \qquad (16)$$

We say that *the restricted Pexider equation* for $f : S_x \rightarrow \mathbb{R}$, $g : S_y \rightarrow \mathbb{R}$, $h : S_{x+y} \rightarrow \mathbb{R}$, that is,

$$h(x+y) = f(x) + g(y) \quad \text{for} \quad (x,y) \in S \ ,$$

can be (uniquely) extended from S to $\mathbb{R}^2$ if there exist (unique) functions $F, G, H : \mathbb{R} \rightarrow \mathbb{R}$ such that

$$\begin{aligned}
F(t) &= f(t) \text{ for } t \in S_x, \\
G(t) &= g(t) \text{ for } t \in S_y, \\
H(t) &= h(t) \text{ for } t \in S_{x+y},
\end{aligned} \qquad (17)$$

and

$$F(x+y) = G(x) + H(y) \quad \text{for all} \quad x \in \mathbb{R}, \ y \in \mathbb{R} \ .$$

We will prove that such *a unique extension of the Pexider equation always exists from a region (nonempty connected open set).* We prove this *first for the hexagonal region*

$$S = S(a,b;r)$$

$$= \{(x,y) \mid x-a, y-b, x+y-a-b \in]-r, r[\} \qquad (18)$$

($a, b \in \mathbb{R}$, $r > 0$). So we have

$$h(x+y) = f(x) + g(y)$$

$$\text{for all } (x,y) \in S(a,b;r) , \tag{19}$$

in particular (since $(a,b) \in S$)

$$h(a+b) = f(a) + g(b) . \tag{20}$$

If we write

$$S^0 = S(0,0;r)$$

$$= \{(u,v) \mid u,v,u+v \in]-r,r[\} \tag{21}$$

then for every $(u,v) \in S^0$ we will have $(u+a,v+b) \in S$. Therefore, by (19),

$$h(u+v+a+b) = f(u+a) + g(v+b)$$

$$\text{for all } (u,v) \in S^0$$

or, after subtracting (20),

$$h(u+v+a+b) - h(a+b) = f(u+a) - f(a)$$

$$+ g(v+b) - g(b) \text{ for all } (u,v) \in S^0 . \tag{22}$$

In particular, if $v = 0$ $(u \in]-r,r[)$ or $u = 0$ $(v \in]-r,r[)$, then

$$h(u+a+b) - h(a+b) = f(u+a) - f(a) \; (u \in]-r,r[)$$

and

$$h(v+a+b) - h(a+b) = g(v+b) - g(b) \; (v \in]-r,r[)$$

respectively. Define now

$$\phi(u) = h(u+a+b) - h(a+b) = f(u+a) - f(a)$$

$$= g(u+b) - g(b) \text{ for all } u \in]-r,r[. \tag{23}$$

By (22), ϕ satisfies the Cauchy equation

$$\phi(u+v) = \phi(u) + \phi(v) \ \text{ for all } \ u,v \in S^0 \qquad (14)$$

that is, when $u,v,u+v \in \,]-r,r[$.

By Corollary 1.8 $(I = \,]-r,r[)$, there exists a unique $\psi : \mathbb{R} \longrightarrow \mathbb{R}$ such that

$$\psi(t) = \phi(t) \ \text{ for } \ t \in \,]-r,r[\qquad (25)$$

and

$$\psi(u+v) = \psi(u) + \psi(v) \ \text{for all } \ (u,v) \in \mathbb{R}^2 \, . \qquad (26)$$

Define now

$$\left.\begin{aligned}
F(x) &= \psi(x-a) + f(a), \\
G(x) &= \psi(x-b) + g(b), \\
H(x) &= \psi(x-a-b) + h(a+b) \ \ (x \in \mathbb{R}).
\end{aligned}\right\} \qquad (27)$$

Then, by (25) and (23), we have (17):

$$F(t) = \phi(t-a) + f(a) = f(t)$$
$$\text{for all } \ t \in \,]a-r, a+r[\, = S_x \, ,$$

$$G(t) = \phi(t-b) + g(b) = g(t)$$
$$\text{for all } \ t \in \,]b-r, b+r[\, = S_y \, ,$$

$$H(t) = \phi(t-a-b) + h(a+b) = h(t)$$
$$\text{for all } \ t \in \,]a+b-r, a+b+r[\, = S_{x+y} \ \ .$$

On the other hand, by (27), (26) and (20),

$$H(x+y) = \psi(x+y-a-b) + h(a+b)$$
$$= \psi(x-a) + \psi(y-b) + f(a) + g(b)$$
$$= F(x) + G(y)$$

for all $x,y \in I\!R$, so F,G,H indeed satisfy the Pexider equation on all of $I\!R^2$. Moreover, from (27), (26) and (20),

$$F(x) = \psi(x) + A, \quad G(x) = \psi(x) + B,$$

$$H(x) = \psi(x) + A + B \quad (x \in I\!R),\qquad (28)$$

$(A = f(a) - \psi(a), \quad B = g(b) - \psi(b))$. The uniqueness of F,G,H follows from the uniqueness of the extension ψ of ϕ (Corollary 1.8 with $I = {]}-r,r{[}$).

Note that $S(a,b;r)$ is a hexagonal neighbourhood of (a,b). On the other hand, *if there exists a unique extension of the Pexider equation from each of two open subsets of $I\!R^2$ with nonempty intersection, then there exists a unique extension also from their union and the three extended triples (F,G,H) are the same, starting from each of these three open sets* (the original two and their union).

Indeed, for the intersection D of the two open sets, $D_x = \{x \mid \exists y : (x,y) \in D\}$ contains a nonempty interval I. Take an arbitrary $(a,b) \in D$ such that $a \in I$. By (23) and (25) the two additive functions ψ_1 and ψ_2 obtained from f starting from the first or second open set, respectively, are *equal on a neighbourhood U of a*. So $\psi_1 - \psi_2$, which is also additive on $I\!R^2$, is 0 on U. (Functions satisfying the Cauchy equation are called '*additive*'.) But, by Corollary 1.5, then $\psi_1(x) - \psi_2(x) = \alpha x$ and $\alpha = 0$, (since $\psi_1 = \psi_2$ on U), so $\psi_1(x) = \psi_2(x)$ *for all* $x \in I\!R$. Also $A = f(a) - \psi(a)$, $B = g(b) - \psi(b)$ are the same for both extensions, and so are F, G, H by (28). — Finally, if R is a region (nonempty open and connected set) then it is path-connected, so a standard compactness argument (we move on a string of intersecting hexagonal neighbourhoods) gives the following.

Theorem 4. *Let $R \subseteq \mathbb{R}^2$ be a region and*

$$R_x = \{x \mid \exists y : (x,y) \in R\},$$

$$R_y = \{y \mid \exists x : (x,y) \in R\}, \qquad (29)$$

$$R_{x+y} = \{x+y \mid (x,y) \in R\} . \qquad (30)$$

If $f : R_x \to \mathbb{R}$, $g : R_y \to \mathbb{R}$ and $h : R_{x+y} \to \mathbb{R}$ satisfy the restricted Pexider equation

$$h(x+y) = f(x) + g(y) \quad for \quad (x,y) \in R , \qquad (31)$$

then there exist unique extensions $F,G,H : \mathbb{R} \to \mathbb{R}$ of f,g,h which satisfy

$$F(x+y) = G(x) + H(y) \quad for \ all \ (x,y) \in \mathbb{R}^2 . \qquad (32)$$

In fact, the general solution of (31) is given by

$$f(t) = \psi(t) + A \quad (t \in R_x),$$

$$g(t) = \psi(t) + B \quad (t \in R_y),$$

$$h(t) = \psi(t) + A + B \quad (t \in R_{x+y}) , \qquad (33)$$

where A,B,C are arbitrary real constants and ψ is an arbitrary solution of the Cauchy equation (26) on $\mathbb{R}^2$.

(The form (33) of the solution follows from (17), (26) and (28) or from (32) and Corollary 2.)

As mentioned before, *a similar theorem holds if $\mathbb{R}^2$ is replaced by the cartesian square of a real or complex linear topological space, with R a nonempty open connected subset of that cartesian square and $f : R_x \to G$, $g : R_y \to G$, $h : R_{x+y} \to G$, where $(G,+)$ is an abelian group.* In particular we have the following.

Corollary 5. *Let* $R \subseteq \mathbb{R}^2_{++}$ *be a region and* $\mathbb{R}^* = \mathbb{R} \backslash \{0\}$ *the set of nonzero reals. Keep the notations (29) and add* $R_{xy} = \{xy \mid (x,y) \in R\}$. *Then the general solutions* $f : R_x \rightarrow \mathbb{R}^*$, $g : R_y \rightarrow \mathbb{R}^*$ *and* $h : R_{xy} \rightarrow \mathbb{R}^*$ *of the restricted Pexider equation*

$$h(xy) = f(x)g(y) \quad for \quad (x,y) \in R \qquad (34)$$

are given by

$$f(t) = aM(t) \ (t \in R_x),$$

$$g(t) = bM(t) \ (t \in R_y),$$

$$h(t) = abM(t) \ (t \in R_{xy}),$$

where $a \in \mathbb{R}^*$, $b \in \mathbb{R}^*$ *are arbitrary constants and* $M : \mathbb{R}_{++} \rightarrow \mathbb{R}_{++}$ *is an arbitrary solution of (12).*

If at least one of f, g, h *is continuous at a point (or bounded on a proper interval or on a set of positive measure, or measurable), then the general solution of (34) is given by*

$$f(t) = at^c \ (t \in R_x),$$

$$g(t) = bt^c \ (t \in R_y),$$

$$h(t) = abt^c \ (t \in R_{xy}), \qquad (35)$$

where $a \in \mathbb{R}^*$, $b \in \mathbb{R}^*$, $c \in \mathbb{R}$ *are arbitrary constants.*

We return now for a moment to the extension of Cauchy's equation (and of multiplicative functions). As we have seen (cf. (14)) such extensions do not always exist, not even from regions. But, of course, every Cauchy equation is a special Pexider equation and so, applying Theorem 4, we have the following (Daróczy-Losonczi [67]).

Corollary 6. *Let $R \subseteq \mathbb{R}^2$ be a region and R_x, R_y, R_{x+y} defined by (29), (30). Then the general solution $f : R_x \cup R_y \cup R_{x+y} \rightarrow \mathbb{R}$ of the restricted Cauchy equation*

$$f(x+y) = f(x) + f(y) \quad for \quad (x,y) \in R$$

is given by

$$f(t) = \begin{cases} \psi(t) + A & for \quad t \in R_x , \\ \psi(t) + B & for \quad t \in R_y , \\ \psi(t) + A + B & for \quad t \in R_{x+y} , \end{cases} \tag{36}$$

where A,B,C are arbitrary real constants and ψ an arbitrary solution of the unrestricted Cauchy equation (26).

The function ψ is called a *quasi-extension* of f. (Of course, (14) is exactly of the form (36) and $\psi(x) = x$ is its quasi-extension.) If $R_x \cap R_y \neq \emptyset$ then $A = B$ in (36), if $R_x \cap R_{x+y} \neq \emptyset$ then $B = 0$, if $R_y \cap R_{x+y} \neq \emptyset$ then $A = 0$ and *if two of these three intersections are nonempty then the quasi-extension ψ is an extension ($A = B = 0$).* - Similar results hold, of course, for the conditional equation for multiplicative functions

$$M(xy) = M(x)M(y) \quad for \quad (x,y) \in R \quad .$$

We will apply now Corollary 5 to so called *'generalized Hicks-neutral production functions'*. A production function describes how the amount of production depends upon several variables ("production factors") one of which may be the time or, in this problem, a parameter describing the state of technology. We will permit the latter variable, denoted by t, to be a completely arbitrary set T (which needs not even to be ordered) and the other variables, $x_1,...,x_n$ to be in open subintervals $I_1,...,I_n$ of $\mathbb{R}_{++}$ (since these 'production factors' may be limited by natural bounds, which certainly need not be the same for all of them).

Hicks [32], Robinson [37] and Gehrig [76] have called a technical invention "neutral" if the ratio of the marginal effect of any two production factors (other than t) remains unchanged as long as the ratio of these two production factors is the same. So *the (partially differentiable) production function $F: I_1 \times ... \times I_n \times T \rightarrow \mathbb{R}$ ($I_1,...,I_n$ open subintervals of $\mathbb{R}_{++}$, T an arbitrary set) is generalized Hicks-neutral if there exist functions $g_{ij}: I_i / I_j \rightarrow \mathbb{R}^*$ such that*

$$\frac{\partial F/\partial x_i}{\partial F/\partial x_j} = g_{ij}(\frac{x_i}{x_j})$$

for all $i \neq j$, $x_k \in I_k$ ($i,j,k=1,...,n$), $t \in T$.   (37)

We will determine here all these F (Aczél-Gehrig [86]). Hicks [32] and Robinson [37] took $n = 2$ (x_1 representing the capital and x_2 the labor) and in this case the invention and the technical process is often called plainly "Hicks-neutral". Interestingly, the case $n > 2$ (but fixed) is easier to solve, so that is what we will do in this section (and the case $n = 2$ in the next). In order that (37) make sense, *the derivatives $\partial F/\partial x_k$ ($k=1,...,n$) have to exist and be nonzero on $I_1 \times ... \times I_n \times T$.* We will first suppose also that they are *continuous,* but will drop this supposition later.

From (37) (written for (i,j), (j,k), then (i,k) — we have at least 3 subscripts to choose from, since $n > 2$), we get

$$g_{ik}\left(\frac{x_i}{x_k}\right) = g_{ij}\left(\frac{x_i}{x_j}\right)g_{jk}\left(\frac{x_j}{x_k}\right)$$

$$(x_i \in I_i, \ x_j \in I_j, \ x_k \in I_k) \ .$$

With $x = x_i/x_j$, $y = x_j/x_k$, this can be written as

$$g_{ij}(xy) = g_{ij}(x)g_{jk}(y)$$

$$(x \in I_{ij} = I_i/I_j, \ y \in I_{jk} = I_j/I_k) \qquad (38)$$

$(I/J = \{u/v \mid u \in I, v \in J\};$ if I, J are open subintervals of $\mathbb{R}_{++}$, so is I/J).

But this is a conditional Pexider equation of the form (34) on the open rectangle $I_{ij} \times I_{jk}$ which, of course, is a region so, by Corollary 5, the general continuous (or just measurable) solutions are given by

$$g_{ij}(x) = a_{ij}x^c \ , \ \ g_{jk}(y) = a_{jk}y^c$$

$$\text{and} \ \ g_{ik}(z) = a_{ij}a_{jk}z^c = a_{ik}z^c \qquad (39)$$

with constant $a_{ij} \neq 0$, $a_{jk} \neq 0$ and c. From the last equation we see that $a_{ij} = a_{ik}/a_{jk}$ or, fixing $k = k_0$ and writing $a_{ik_0} = a_i$, $a_{jk_0} = a_j$,

$$a_{ij} = \frac{a_i}{a_j} \ .$$

So (37) becomes

$$\frac{\partial F/\partial x_i}{\partial F/\partial x_j} = \frac{a_i x_i^b}{a_j x_j^b} \quad (i,j=1,2,...,n) \ . \qquad (40)$$

(We did not exclude $i = j$, because (40) is true then too, though trivial). But this can be written as

$$\begin{vmatrix} \dfrac{\partial F}{\partial x_i} & \dfrac{\partial F}{\partial x_j} \\[2ex] \dfrac{\partial \Phi_{i,j}}{\partial x_i} & \dfrac{\partial \Phi_{i,j}}{\partial x_j} \end{vmatrix} = 0 , \tag{41}$$

where

$$\Phi_{i,j}(x_i, x_j) = \alpha_i x_i^{\beta} + \alpha_j x_j^{\beta}$$

$$\text{if } b \neq -1 \;\; (\beta = b+1, \alpha_k = \frac{a_k}{\beta}) \tag{42}$$

and

$$\Phi_{i,j}(x_i, x_j) = a_i \log x_i + a_j \log x_j$$

$$\text{if } b = -1 . \tag{43}$$

It is known (see e.g. Courant-John [74], Marsden [74]) that, if on a region (in our case on $I_i \times I_j$)

$$\begin{vmatrix} \dfrac{\partial F}{\partial x} & \dfrac{\partial F}{\partial y} \\[2ex] \dfrac{\partial \Phi}{\partial x} & \dfrac{\partial \Phi}{\partial y} \end{vmatrix} = 0 \;\; \text{and} \;\; \frac{\partial \Phi}{\partial x} \neq 0 ,$$

then there exists a differentiable one-place function G such that

$$F(x,y) = G[\Phi(x,y)]$$

on that region. In our case there are variables in F, other than x_i and x_j (t and at least one more), which we have kept constant when forming the partial derivatives $\partial F/\partial x_i$, $\partial F/\partial x_j$, so that (41), for instance with $i = 1$, $j = 2$, implies only

$$F(x_1,x_2,x_3,...,x_n,t) \;=\; G\big(\alpha_1 x_1^{\beta}+\alpha_2 x_2^{\beta},x_3,...,x_n,t\big) \qquad (44)$$

resp.

$$F(x_1,x_2,x_3,...,x_n,t) \;=\; G\big(a_1\log x_1+a_2\log x_2,x_3,...,x_n,t\big) \;,\; (45)$$

according to whether we had (42) or (43). (One sees from (40), (42) and (43) that b cannot depend upon $x_3,...,x_n,t$ and that a_1,a_2 or α_1,α_2 may depend on them at most in the form of a common factor which can be immersed into the dependence of G upon $x_3,...,x_n,t$.) We see also that G has nonzero derivatives in all variables but t.

Let us take, for instance, $b = -1$, that is, (43) and (45) ((42) and (44) can be handled in the same way). We use now (40) with $i = 1$, $j = 3$ and get, in view of (45) and with the notation $u = a_1 \log x_1 + a_2 \log x_2$,

$$\frac{(\partial G/\partial u)a_1 x_1^{-1}}{\partial G/\partial x_3} \;=\; \frac{a_1 x_1^{-1}}{a_3 x_3^{-1}} \;,$$

which is again of the form

$$\begin{vmatrix} \dfrac{\partial G}{\partial u} & \dfrac{\partial G}{\partial x_3} \\[2mm] \dfrac{\partial \Psi}{\partial u} & \dfrac{\partial \Psi}{\partial x_3} \end{vmatrix} = 0 \;\; \text{with} \;\; \Psi(u,x_3) = u + a_3 \log x_3 \;.$$

Therefore and because $\partial \Psi/\partial u = 1 \neq 0$, there exists now a function H, differentiable (with nonzero derivatives) in all variables but t, such that

$$G(u,x_3,...,x_n,t) = H(u+a_3 \log x_3,x_4,...,x_n,t) \;,$$

or, in view of (45),

$$F(x_1, x_2, x_3, x_4, ..., x_n, t)$$

$$= H(a_1 \log x_1 + a_2 \log x_2 + a_3 \log x_3, x_4, ..., x_n, t) \ .$$

Clearly we can continue in the same way and get

$$F(x_1, ..., x_n, t) = \phi(a_1 \log x_1 + ... + a_n \log x_n, t)$$

$$(x_k \in I_k; \ k=1, ..., n; \ t \in T) \tag{46}$$

in the case (43), (45), while in the case (42), (44) we get

$$F(x_1, ..., x_n, t) = \phi(\alpha_1 x_1^\beta + ... + \alpha_n x_n^\beta, t)$$

$$(x_k \in I_k; \ k=1, ..., n; \ t \in T) \tag{47}$$

$(\beta \neq 0)$. — Conversely, both (46) and (47) satisfy (37) if ϕ has a nonzero derivative in the first variable and $a_1, ..., a_n, \alpha_1, ..., \alpha_n, \ \beta$ are nonzero constants. — Production functions of the form $(\sum_{k=1}^{n} \alpha_k x_k^\beta)^{1/\beta}$ or

$$\exp(\sum_{k=1}^{n} a_k \log x_k / \sum_{k=1}^{n} a_k) = \prod_{k=1}^{n} x_k^{q_k} \ (q_k \neq 0, \sum_{k=1}^{n} q_k = 1)$$

are called ACMS (Arrow-Chenery-Minhas-Solow) and Cobb-Douglas functions, respectively (see Eichhorn [78]). So we have proved the following.

Theorem 7. *Let T be an arbitrary set, $I_k \subseteq \mathbb{R}_{++}$ $(k=1, ..., n; \ n > 2)$ open intervals and suppose that the function $F: I_1 \times ... \times I_n \times T \longrightarrow \mathbb{R}$ has continuous nonzero first partial derivatives with respect to its first n variables. Then the general solutions of (37) are given by (46) and (47), where ϕ is an arbitrary two-place function with continuous nonzero derivative in its first variable and $a_1, ..., a_n, \alpha_1, ..., \alpha_n, \beta$ are arbitrary nonzero constants.*

So all generalized Hicks-neutral production functions

are t-dependent functions either of ACMS or of Cobb-Douglas functions or of $\prod\limits_{k=1}^{n} x_k^{a_k}$ *with* $\sum\limits_{k=1}^{n} a_k = 0.$

(Because, if $\sum\limits_{k=1}^{n} a_k \neq 0$ then $\prod\limits_{k=1}^{n} x_k^{a_k} = (\prod\limits_{k=1}^{n} x_k^{a_k/\sum a_k})^{\sum a_k}$.)

As we mentioned above, *the condition of the continuity of the partial derivatives of F can be eliminated.* We needed it only to solve the (conditional) Pexider equation (38) and get (39) as solutions. But (37) gives with $x_i = x$, $x_k = a_k$ (constants) for $k \neq i$,

$$g_{ij}(\frac{x}{a_j}) = \frac{\partial F/\partial x_i}{\partial F/\partial x_j} \,(a_1,...,a_{i-1},x,a_{i+1},...,a_n,t) =$$

$$\lim_{m \to \infty} \frac{F(a_1,...,a_{i-1},x+\dfrac{1}{m},a_{i+1},...,a_n,t)-F(a_1,...,a_{i-1},x,a_{i+1},...,a_n,t)}{F(a_1,...,x,...,a_{j-1},a_j+\dfrac{1}{m},a_{j+1},...,a_n,t)-F(a_1,...,x,...,a_{j-1},a_j,a_{j+1},...,a_n,t)}$$

(without loss of generality we may suppose that the denominator does not vanish for any finite m). The function after the lim sign is continuous in x for all m, so the limit, though not necessarily continuous, is *measurable*. But, cf. Corollary 5, the measurability of g_{ij} is sufficient to make (39) the general solution of (38) and then everything goes on as before. (Of course, the derivative of ϕ with respect to its first variable needs not to be continuous anymore.)

Often $I_1 =...= I_n = \mathbb{R}_{++}$ is chosen and *first degree homogeneity* of F in the first n variables, that is,

$$F(x_1 u,...,x_n u,t) = uF(x_1,...,x_n,t)$$

$$(x_1,...,x_n, u \in \mathbb{R}_{++}, t \in T) \qquad (48)$$

is also supposed. Then for (46) which, as we have seen, can be written either as

$$F(x_1,...,x_n,t) = \psi(\prod_{k=1}^{n} x_k^{q_k},t)$$

$$(q_k \neq 0;\ k=1,...,n;\ \sum_{k=1}^{n} q_k = 1)\ ,\qquad\qquad (49)$$

or as

$$F(x_1,...,x_n,t) = \psi(\prod_{k=1}^{n} x_k^{a_k},t)$$

$$(a_k \neq 0;\ k=1,...,n;\ \sum_{k=1}^{n} a_k = 0)\ ,\qquad\qquad (50)$$

we have in the case (49)

$$\psi(u\prod_{k=1}^{n} x_k^{q_k},t) = F(x_1 u,...,x_n u,t)$$

$$= F(x_1,...,x_n,t)u = \psi(\prod_{k=1}^{n} x_k^{q_k},t)u$$

that is,

$$\psi(ux,t) = \psi(x,t)u\ \ .$$

With

$$u = \frac{1}{x}\ ,\ h(t) = \psi(1,t)$$

this gives

$$\psi(x,t) = h(t)x \qquad\qquad (51)$$

and (49) becomes

$$F(x_1,...,x_n,t) = \prod_{k=1}^{n} x_k^{q_k} h(t)$$

$$(x_1,...,x_n \in \mathbb{R}_{++},\ t \in T) \qquad\qquad (52)$$

$(q_k \neq 0, \sum\limits_{k=1}^{n} q_k = 1)$. The form (50) is incompatible with (48):

$$\psi(\prod_{k=1}^{n} x_k^{a_k}, t) = F(x_1 u, \ldots, x_n u, t)$$

$$= F(x_1, \ldots, x_n, t)u = \psi(\prod_{k=1}^{n} x_k^{a_k}, t)u$$

would imply $\psi(x, t) \equiv 0$ so F (and its derivatives) would be identically zero.

For (47), which can be written as

$$F(x_1, \ldots, x_n, t) = \psi((\sum_{k=1}^{n} \alpha_k x_k^{\beta})^{1/\beta}, t) \quad (\beta \neq 0) , \qquad (53)$$

equation (48) means

$$\psi(u(\sum_{k=1}^{n} \alpha_k x_k^{\beta})^{1/\beta}, t) = \psi((\sum_{k=1}^{n} \alpha_k x_k^{\beta})^{1/\beta}, t)u$$

so we have (51) again and, from (53),

$$F(x_1, \ldots, x_n, t) = (\sum_{k=1}^{n} \alpha_k x_k^{\beta})^{1/\beta} h(t)$$

$$(x_1, \ldots, x_n \in \mathbb{R}_{++}, t \in T) . \qquad (54)$$

Of course no values of h should be 0. We may suppose that h is *positive-valued*. If F is supposed to be *increasing* in its first n variables, then the $q_1, \ldots, q_n$ in (52) and the $\alpha_1, \ldots, \alpha_n$ in (54) are *positive* (while (50) is *never increasing* in all x_k). (The same holds in (49), (53) and (50) *without homogeneity* if ψ and F are *increasing* in their first or first n variables, respectively.) So we have the following.

Corollary 8. *The general solution* $F: \mathbb{R}^n_{++} \times T \to \mathbb{R}$ *(n > 2, T an arbitrary set) of*

$$\frac{\partial F/\partial x_i}{\partial F/\partial x_j} = g_{ij}\left(\frac{x_i}{x_j}\right)$$

$$(i \neq j;\ x_k \in \mathbb{R}_{++};\ i,j,k=1,...,n)\quad,$$

which is homogeneous of first degree in the first n variables and has positive first derivatives in these variables, is given by (52) and (54), where $h: T \to \mathbb{R}_{++}$ *is an arbitrary function,* $\beta \in \mathbb{R}^*$, $\alpha_k \in \mathbb{R}_{++}$, $q_k \in \mathbb{R}_{++}$ $(k=1,...,n)$ *are arbitrary constants except that* $\sum_{k=1}^{n} q_k = 1$. *So all generalized Hicks-neutral production functions are ACMS or Cobb-Douglas functions multiplied by an arbitrary function of the time (or technology) parameter.*

Also $\partial^2 F/\partial x_k^2 < 0$ is often supposed. *If this is satisfied at least for one* k, *in addition to the above, then* $\beta < 1$.

Such results were obtained previously (see Gehrig [76]) under further suppositions on the first and second derivatives. There the results for $n > 2$ were derived from the solution in the case $n = 2$. But we will see in the next section that the solution for $n = 2$ is rather more difficult and lengthier than for $n > 2$.

6. Determination of all Hicks-neutral production functions depending upon capital, labor and time (state of technology). The translation equation.

We come now to the case $n = 2$, the Hicks-neutral production processes originally envisaged by Hicks [32], Robinson [37], and Sato-Beckmann [68], who took capital and labor (in addition to the time or state of technology) as production factors. As we will see, homogeneity will play an important role here too, as it did in section 3 (see, for instance, Proposition 3.4). We will prove the following (cf. Aczél-Gehrig [86], where an arbitrary open rectangle $I_1 \times I_2$ stands in place of $\mathbb{R}^2_{++}$, just as in Theorem 5.7 above).

Theorem 1. *Let T be an arbitrary set and suppose that the function $F: \mathbb{R}^2_{++} \times T \to \mathbb{R}$ has nonzero first partial derivatives with respect to its first two variables. Then all solutions of*

$$\frac{\partial F/\partial x_1}{\partial F/\partial x_2} = g\left(\frac{x_1}{x_2}\right) \quad (x_1, x_2 \in \mathbb{R}_{++}, \, t \in T) \qquad (1)$$

are of the following form

$$F(x_1, x_2, t) = \psi_{(i)}[\Phi_{(i)}(x_1, x_2), t]$$

$$\left(x_1, x_2 \in \mathbb{R}_{++}, \, \frac{x_1}{x_2} \in I_i, \, t \in T\right), \qquad (2)$$

where the $\psi_{(i)}$ are two-place functions with nonzero derivative in their first variable, i is in an at most countable *index set, the I_i's are disjoint open subintervals of $\mathbb{R}_{++}$ and the $\Phi_{(i)}$ have nonzero derivatives and satisfy*

$$\Phi_{(i)}(x_1 u, x_2 u) = \Phi_{(i)}(x_1, x_2) u$$

$$\text{whenever } \frac{x_1}{x_2} \in I_i \text{ and } u \in \mathbb{R}_{++} \quad (3_1)$$

(that is, the $\Phi_{(i)}$ are homogeneous of first degree), while

$$F(x_1 u, x_2 u, t) = F(x_1, x_2, t)$$

$$\text{whenever } \frac{x_1}{x_2} \in \mathbb{R}_{++} \setminus \cup I_i, \ u \in \mathbb{R}_{++}, \ t \in T \quad (3_0)$$

(that is, F is homogeneous of zeroth degree in its first two variables).

(The $\psi_{(i)}$ and $\Phi_{(i)}$ should be adjusted at the boundaries of their respective domains so that F has (nonzero) partial derivatives.)

Note that for $n = 2$ we had in (5.49) and (5.53) t-dependent functions of *special* homogeneous functions of first degree, namely of $\prod_{k=1}^{n} x_k^{q_k}$ $(q_k \neq 0, \sum_{k=1}^{n} q_k = 1)$ or of $(\sum_{k=1}^{n} \alpha_k x_k^{\beta})^{1/\beta}$ $(\alpha_k \neq 0, \beta \neq 0)$, respectively, while in (5.50) we had a t-dependent function of a special homogeneous function of zeroth degree, namely of $\prod_{k=1}^{n} x_k^{a_k}$ $(a_k \neq 0, \sum_{k=1}^{n} a_k = 0)$. For $n = 2$, as (2)-(3_1) and (3_0) show, t-dependent functions of *any* homogeneous function of first or zeroth degree and their combinations with nonzero partial derivatives are solutions. Further restrictions will eliminate homogeneous functions of zeroth degree (3_0) (remember that also (5.50) was eliminated) and restrict somewhat ψ and the first degree homogeneous function Φ (there will then be just one ψ and one Φ).

Proof of Theorem 1. We have seen in the proof of Theorem 5.7 how the parameter t is dragged along through the proof without having any essential role. In order to simplify the present proof, *we ignore the parameter t till the end.* (The reader may insert it in every formula if this helps understanding.) We will also denote simply by F'_1 and F'_2 the first derivative of F with respect to the first or second variable, respectively. So we get from (1)

$$\frac{F'_1(x_1,x_2)}{F'_2(x_1,x_2)} = g(\frac{x_1}{x_2}) = g(\frac{x_1 u}{x_2 u})$$

$$= \frac{F'_1(x_1 u,x_2 u)u}{F'_2(x_1 u,x_2 u)u} = \frac{H'_1(x_1,x_2,u)}{H'_2(x_1,x_2,u)} , \qquad (4)$$

where

$$H(x_1,x_2,u) = F(x_1 u,x_2 u) \quad (x_1,x_2,u \in \mathbb{R}_{++}) .$$

Indeed then

$$H'_j(x_1,x_2,u) = \frac{\partial}{\partial x_j} F(x_1 u,x_2 u)$$

$$= F'_j(x_1 u,x_2 u) \frac{\partial (x_j u)}{\partial x_j}$$

$$= F'_j(x_1 u,x_2 u)u \quad (j=1,2) .$$

The two extremities of (4) give an equation of the same form as (5.41),

$$\begin{vmatrix} H'_1 & H'_2 \\ F'_1 & F'_2 \end{vmatrix} = 0 ,$$

this time on the region $\mathbb{R}^2_{++}$ and, since $F'_1 \neq 0$ on the same region, we have

$$F(x_1 u, x_2 u) = H(x_1, x_2, u) = G[F(x_1, x_2), u]$$

$$(x_1, x_2, u \in \mathbb{R}_{++}) \ . \tag{5}$$

Let $I = F(\mathbb{R}_{++}, \mathbb{R}_{++})$ be *the image of* $\mathbb{R}^2_{++}$ *under* F. Since F is differentiable with nonzero derivatives, I is an (open) interval. As before, $G : I \times \mathbb{R}_{++} \to I$ in (5) is differentiable in both variables, the derivative with respect to the first variable being nonzero, since

$$G'_1[F(x_1, x_2), u] F'_1(x_1, x_2) = \frac{\partial}{\partial x_1} F(x_1 u, x_2 u)$$

$$= F'_1(x_1 u, x_2 u) u \neq 0 \ ;$$

(it does not follow that $G'_2 \neq 0$, since

$$0 = G'_2[F(x_1, x_2), u] = F'_1(x_1 u, x_2 u) x_1 + F'_2(x_1 u, x_2 u) x_2$$

does not contradict $F'_1 \neq 0$, $F'_2 \neq 0$). Repeated application of (5) gives

$$G[F(x_1, x_2), uv] = F(x_1 uv, x_2 uv)$$

$$= G[F(x_1 u, x_2 u), v]$$

$$= G(G[F(x_1, x_2), u], v)$$

that is,

$$G(x, uv) = G[G(x, u), v]$$

$$\text{for all} \quad u, v \in \mathbb{R}_{++}, \quad x \in I \ . \tag{6}$$

This is called the *multiplicative translation equation*. We will see later (Theorem 3) that *the general solution of (6), continuous in each variable and nonconstant in the second* is given by

$$G(x,u) = \psi[\psi^{-1}(x)u] \quad (x \in I, \; u \in \mathbb{R}_{++}) \qquad (7)$$

where $\psi : \mathbb{R}_{++} \to I$ is an arbitrary continuous and strictly monotonic function. (In our case, since G has a nonzero partial derivative, also ψ' exists with respect to x, and is nowhere 0). We will see, by the way, that under these conditions I *has to be open*, even if we had not known this in advance. Moreover, from (7),

$$G(x,1) = x \quad \text{for all} \; x \in I \; . \qquad (8)$$

This follows also *from (5)*, by putting there $u = 1$, $F(x_1,x_2) = x$. The latter fact is of some importance *if G is constant in its second variable* (so one of the conditions of Theorem 3 below is *not* satisfied). Then

$$G(x,u) = h(x) \quad \text{for all} \; u \in \mathbb{R}_{++}$$

(notice that here $G_2'(x,u) \equiv 0$) and by (8) $h(x) = x$, so

$$G(x,u) = x \quad \text{for all} \; u \in \mathbb{R}_{++} \; . \qquad (9)$$

Of course, it can well be that $u \mapsto G(x,u)$ is constant only for some x in I. (For what follows, cf. Sibirsky [75] and Förg-Rob [86].) If the set of such x is dense in a subinterval I_0 of I then, since G is continuous in x, (9) has to hold for all I_0, even for its closure $\bar{I}_0$. Let us extend $\bar{I}_0$ as far as possible, so that

$$G(x,u) = x \quad \text{for all} \; x \in I', \; u \in \mathbb{R}_{++} \; . \qquad (10)$$

Of course, also I' will be a *closed interval* except if an endpoint of I' is also an endpoint of I. If $I' = I$, then (9) holds for all $x \in I$. If the set of x for which $u \mapsto G(x,u)$ is constant is *not everywhere dense* in I then there exist proper intervals I'' such that $u \mapsto G(x,u)$ is not constant for any $x \in I''$. So I splits into closed or half-closed intervals (or points) I' where (10) holds and into open intervals I'' where (cf. (7))

$$G(x,u) = \psi[\psi^{-1}(x)u]$$

$$\text{for all } x \in I'', \ u \in \mathbb{R}_{++}. \tag{11}$$

Of course, there may be several such I' and I'' (with different ψ).

If $x = F(x_1,x_2)$ is in an I' then, by (5),

$$F(x_1u,x_2u) = F(x_1,x_2)$$

$$(u \in \mathbb{R}_{++}, \ F(x_1,x_2) \in I') \ , \tag{12}$$

that is, *we have* (3_0). We see that, whenever (12) holds for (x_1,x_2), it holds on the whole *ray* $\{(x_1v,x_2v) \mid v \in \mathbb{R}_{++}\}$ going through (x_1,x_2) and the origin. If, on the other hand, $x = F(x_1,x_2)$ is in an I'' then, again by (5),

$$F(x_1u,x_2u) = \psi(\psi^{-1}[F(x_1,x_2)]u) \ .$$

So the new function Φ, defined by

$$\Phi(x_1,x_2) = \psi^{-1}[F(x_1,x_2)]$$

$$(F(x_1,x_2) \in I'') \ , \tag{13}$$

(which also has nonzero partial derivatives) will satisfy

$$\Phi(x_1u,x_2u) = \Phi(x_1,x_2)u$$

$$(u \in \mathbb{R}_{++}, \ \psi[\Phi(x_1,x_2)] \in I'') \ . \tag{14}$$

Again, if (14) holds for an (x_1,x_2), it holds on the whole ray going through (x_1,x_2) and the origin. So we have (13) with (14) on *sectors* $\{(x_1,x_2) \mid (x_1/x_2) \in I_i\}$ that is, *we have (2) with* (3_1). Since the intervals I_i are open, each is of positive length so there are at most countably many of them. The components of the complement $\mathbb{R}_{++} \backslash \cup I_i$ may be points, proper closed intervals, and possibly, half open intervals 'at the ends'. These correspond to the I' above.   □

The *converse*, that (2) with (3_1), where $\psi'_{(i)}$, $\Phi'_{(i)1}$, and $\Phi'_{(i)2}$ exist and are different from 0, satisfies (1) and has nonzero first derivatives, in each sector, is obvious using (2) and (3_1) (we omit the subscripts (i)):

$$F'_1(x_1,x_2,t) = \psi'_1[\Phi(x_1,x_2),t]\,\Phi'_1(x_1,x_2) \neq 0 \ ,$$

$$F'_2(x_1,x_2,t) = \psi'_1[\Phi(x_1,x_2),t]\,\Phi'_2(x_1,x_2) \neq 0 \ ,$$

$$\frac{F'_1(x_1,x_2,t)}{F'_2(x_1,x_2,t)} = \frac{\Phi'_1(x_1,x_2)u}{\Phi'_2(x_1,x_2)u}$$

$$= \frac{\Phi'_1(x_1u,x_2u)u}{\Phi'_2(x_1u,x_2u)u}$$

$$= \frac{\Phi'_1(x_1/x_2,1)}{\Phi'_2(x_1/x_2,1)} = g\left(\frac{x_1}{x_2}\right) \ .$$

(We have differentiated with respect to x_1 and to x_2 and divided, then substituted $u = 1/x_2$). Similarly, from (3_0) with $u = 1/x_2$,

$$F(x_1,x_2,t) = F(x_1u,x_2u,t)$$

$$= F(\frac{x_1}{x_2},1,t) = \Psi(\frac{x_1}{x_2},t),$$

$$\frac{F_1'(x_1,x_2,t)}{F_2'(x_1,x_2,t)} = \frac{\Psi_1'(x_1/x_2,t)/x_2}{\Psi_1'(x_1/x_2,t)(-x_1/x_2^2)}$$

$$= -\left|\frac{x_1}{x_2}\right|^{-1} .$$

We see that all two-place homogeneous functions of zeroth degree on $\mathbb{R}^2_{++}$ are functions of $\frac{x_1}{x_2}$. Similarly, all two-place homogeneous functions of first degree on $\mathbb{R}^2_{++}$ are of the form $x_2 f(\frac{x_1}{x_2})$. For instance in the latter case, again with $u = 1/x_2$,

$$\Phi(x_1,x_2) = \frac{1}{u} \Phi(x_1 u, x_2 u) = x_2 \Phi(\frac{x_1}{x_2},1) .$$

So the parts (3_0) and (2)-(3_1) of the solutions of (1) can be written as

$$F(x_1,x_2,t) = \phi(\frac{x_1}{x_2},t) \tag{15}$$

and

$$F(x_1,x_2)t = \psi[x_2 f(\frac{x_1}{x_2}),t] , \tag{16}$$

respectively. The formula (15) certainly does not look right for our purposes since it could mean that the production increases with decreasing labor. While surprisingly something like this has been said at some time or other both by British conservatives (with regard to the number of workers) and by German trade unionists (regarding the working time per worker), we don't want to stretch this point. We only point out that the same thing can happen also with (16), for instance for $f(v) = v^2$, and that, on the other hand, both (for instance (15) with $\phi(v,t) = h(t)/v$, (16) with $\psi(x,t) = xh(t)$, $f(v) = 1/v$) can give production functions decreasing with the amount of capital (throwing money at problems may make them worse). On the other hand, these observations could furnish the motive to suppose right away that F is *increasing* in both x_1 and x_2 (production increases if capital and/or labor increases). This *eliminates the solutions (15)* (that is, (3_0)) immediately and also the above 'bad' examples to (16) (or to (2) with (3_1)); cf. Remark after Corollary 2.

Another way to *eliminate (15)* is to suppose again (as in Corollary 5.8) that F *is homogeneous of first degree* in x_1, x_2:

$$F(x_1 u, x_2 u, t) = uF(x_1, x_2, t)$$

$$(x_1, x_2 \in \mathbb{R}_{++}, \; t \in T) \; . \tag{17}$$

Indeed,   (15)   (or (3_0))   gives   of   course $F(x_1 u, x_2 u, t) = F(x_1, x_2, t)$ which contradicts (17) (since $F \not\equiv 0$). But *(17) restricts also* (2)-(3_1) (that is, (16)): Then

$$u\psi[\Phi(x_1, x_2), t] = uF(x_1, x_2, t) = F(ux_1, ux_2, t)$$

$$= \psi[\Phi(ux_1, ux_2), t]$$

$$= \psi[u\Phi(x_1, x_2), t] \; .$$

So

$$u\psi(x,t) = \psi(xu, t)$$

and, with $u = 1/x$,

$$\psi(x,t) = x\psi(1,t) = xh(t)$$

so that (2) becomes

$$F(x_1,x_2,t) = \Phi(x_1,x_2)h(t)$$

and (16) becomes

$$F(x_1,x_2,t) = x_2 f(\frac{x_1}{x_2})h(t)$$

$$(x_1,x_2 \in \mathbb{R}_{++},\ t \in T) \quad . \tag{18}$$

We again may suppose that h is *positive-valued*.

If we have also that F is increasing, then the nonzero F_1' and F_2' have to be *positive*, so from (18),

$$f'(\frac{x_1}{x_2}) > 0 \ , \ \ f(\frac{x_1}{x_2}) - \frac{x_1}{x_2}\, f'(\frac{x_1}{x_2}) > 0 \ \ .$$

We have proved the following.

Corollary 2. *The general solution* $F : \mathbb{R}^2_{++} \times T \to \mathbb{R}$ *(T an arbitrary set) of (1), which is homogeneous of first degree in the first two variables and has positive first derivatives in these variables is given by (18) where* $h : T \to \mathbb{R}_{++}$ *is an arbitrary function and* $f : \mathbb{R}_{++} \to \mathbb{R}$ *satisfies*

$$f(x) > xf'(x) > 0 \quad (x \in \mathbb{R}_{++}) \tag{19}$$

but is otherwise arbitrary.

Remark. We get *the same conditions (19) for f in (16) if* homogeneity of F is not supposed but, *in addition to* $F_1' > 0$, $F_2' > 0$, *we have also* $\psi_1' > 0$, which may be assumed without loss of generality, since ψ is strictly monotonic in its first variable and $\psi_1' \neq 0$.

Often also $F_{11}'' = \partial^2 F/\partial x_1^2 < 0$, $F_{22}'' = \partial^2 F/\partial x_2^2 < 0$ are supposed (Sato-Beckmann [68]). *Either of these is equivalent to* $f'' < 0$ *in (18) (with* $h > 0$) *or in (16) (with* $\psi_1' > 0$).

Now we come to the promised result about (6). In order to make the calculations easier, we transform (6): with $u = e^s$, $v = e^t$ (s,t arbitrary real numbers) and

$$\Psi(x,s) = G(x,e^s) \quad (x \in I, \ s \in \mathbb{R}), \tag{20}$$

equation (6) goes over into the (additive) *translation equation*

$$\Psi(x,s+t) = \Psi[\Psi(x,s),t] \quad (x \in I; \ s,t \in \mathbb{R}) \ . \tag{21}$$

We prove the following (see e.g. Aczél [66]).

Theorem 3. *The general solution* $\Psi: I \times \mathbb{R} \to I$ (*I a real interval*) *of (21), continuous in both variables and noncon-stant in the second (that is,* $u \mapsto \Psi(x,u)$ *is nonconstant for every* $x \in I$) *is given by*

$$\Psi(x,s) = \lambda[\lambda^{-1}(x)+s] \quad (x \in I, \ s \in \mathbb{R}), \tag{22}$$

where $\lambda: \mathbb{R} \to I$ *is an arbitrary continuous strictly mono-tonic function.*

Under these conditions the interval I has to be open.

From (22) we get (7) immediately: (20) *with* $\psi(t) = \lambda(\log t)$, $\psi^{-1}(x) = e^{\lambda^{-1}(x)}$ *gives*

$$G(x,u) = \Psi(x,\log u) = \lambda[\lambda^{-1}(x)+\log u]$$
$$= \psi(e^{\lambda^{-1}(x)+\log u}) = \psi[\psi^{-1}(x)u]$$
$$(x \in I, \ u \in \mathbb{R}_{++})$$

as asserted.

Proof of Theorem 3. It is clear that (21) is satisfied by (22) with arbitrary continuous and strictly monotonic λ and that the Ψ defined by (22) with such λ are continuous and nonconstant (even strictly monotonic) in both variables.

Conversely, put $x = x_0$ into (21) and define $\lambda: \mathbb{R} \to I$ by

$$\lambda(t) = \Psi(x_0, t) \quad (t \in \mathbb{R}) \tag{23}$$

in order to get

$$\lambda(s + t) = \Psi[\lambda(s), t] \quad (s, t \in \mathbb{R}) . \tag{24}$$

This leads to (22) if we prove (i) that λ, which by (23) is continuous, is also strictly monotonic and (ii) that λ takes as values all points of I ($\lambda(\mathbb{R}) = I$).

(i) If the continuous λ were not strictly monotonic, then there would exist two reals $t_1 \neq t_2$ even as close to each other as we want to (given ϵ, $0 < |t_1 - t_2| < \epsilon$) such that $\lambda(t_1) = \lambda(t_2)$. So, by (24),

$$\lambda[t + (t_1 - t_2)] = \lambda[t_1 + (t - t_2)] = \Psi[\lambda(t_1), t - t_2]$$
$$= \Psi[\lambda(t_2), t - t_2] = \lambda(t_2 + t - t_2) = \lambda(t) ,$$

that is, λ would be periodic with *arbitrarily small periods* so, being continuous, it would be constant, contrary to supposition. Therefore λ is indeed strictly monotonic.

(ii) Let λ be, for instance, strictly increasing and b the right endpoint of I. We prove that

$$\lim_{t \to \infty} \lambda(t) = b .$$

(The proof of

$$\lim_{t \to -\infty} \lambda(t) = a,$$

where a is the left endpoint of I, is similar and the proofs for decreasing λ differ very little.) To be exact, we prove that

$$b' = \lim_{t \to \infty} \lambda(t)$$

cannot be in I. Since λ is increasing, this will indeed show that $b' = b$ and also that $b \notin I$, so I is *open* from above (and, similarly, from below). If we had $b' \in I$, then $s \mapsto \Psi(b',s)$ could not be constant either, so there would exist two numbers, s_1 and s_2, such that $\Psi(b',s_1) \neq \Psi(b',s_2)$ and, Ψ being continuous in the first variable,

$$b' = \lim_{t \to \infty} \lambda(t+s_1) = \lim_{t \to \infty} \Psi[\lambda(t),s_1]$$

$$= \Psi(b',s_1) \neq \Psi(b',s_2) = \lim_{t \to \infty} \Psi[\lambda(t),s_2]$$

$$= \lim_{t \to \infty} \lambda(t+s_2) = b'$$

which is impossible.

Thus for every $x \in I$ there exists a (unique) $s \in \mathbb{R}$ such that

$$x = \lambda(\dot{s}), \; s = \lambda^{-1}(x)$$

and (24) goes over into (22).   □

To conclude this section we mention that the solution of (6) would have given also another method to solve (2.2). Indeed

$$u(rx) - u(ry) = G[u(x)-u(y),r]$$

implies

$$G[u(x)-u(y),rs] = u(rsx)-u(rsy)$$

$$= G[u(rx)-u(ry),s] = G(G[u(x)-u(y),r],s)$$

that is,

$$G(z,rs) = G[G(z,r),s] \quad (z \in I,\ s \in \mathbb{R}_{++}) \quad ,$$

which is exactly (6).

7. The associativity equation. Synthesis of ratio judgements. The quasiarithmetic means. The Jensen equations. A conditional linear-affine equation. A characterization of root-mean-powers and of the geometric mean.

While we have often dealt with functional equations for multiplace functions in the present work, the translation equations (6.6) and (6.21) are different: They are *composite equations*: the unknown function appears again inside the unknown function (on the right hand side). Here we will solve another important composite equation

$$F[F(x,y),z] = F[x,F(y,z)]$$

$$(x,y,z \in I, \ F:I^2 \longrightarrow I) \qquad (1)$$

(where $I \subseteq R$ is a proper interval) and apply it to a problem of *synthesizing judgements*, which is somewhat similar to the allocation problem with which we have started section 1. (So, in a way, we have run a full circle). Equation (1) is called the *associativity equation*, for obvious reasons which become clear if we write (1) as

$$(x \circ y) \circ z = x \circ (y \circ z)$$

$$\text{for all} \ \ x,y,z \in I. \qquad (2)$$

We will prove the following.

Theorem 1. *Let I be a (closed, open, half-open, finite or infinite) proper interval of real numbers. Then $\circ: I^2 \to I$ is a continuous operation on I which satisfies (2) and is cancellative, that is,*

$$t_1 \circ z = t_2 \circ z \ or \ z \circ t_1 = z \circ t_2$$

$$imply \ t_1 = t_2 \ for \ any \ z \in I$$

if, and only if, there exists a continuous and strictly monotonic function $\phi: J \to I$ such that

$$x \circ y = \phi[\phi^{-1}(x) + \phi^{-1}(y)]$$

$$for \ all \ \ x, y \in I \ . \tag{3}$$

Here J is one of the real intervals

$$]-\infty, \gamma], \ ]-\infty, \gamma[, \ [\delta, \infty[, \ ]\delta, \infty[$$

$$or \ \ ]-\infty, \infty[\ = \mathbb{R} \tag{4}$$

for some $\gamma \leq 0 \leq \delta$. Accordingly I has to be open at least from one side.

The function ϕ in (3) is unique up to a linear transformation of the variable ($\phi(x)$ may be replaced by $\phi(Cx)$, $C \neq 0$ but by no other function).

Remark. Theorem 1 is equivalent to the statement that *every continuous cancellative real semigroup is isomorphic in an order-preserving and continuous way (o-iseomorphic) to one of the intervals (4) under ordinary addition* (Aczél [66]). Note that $\circ$ was not supposed to be commutative (symmetric) but the result (3) shows that it is.

Proof of Theorem 1. The 'if' part is obvious. We prove the 'only if' part in several steps.

(o) *The operation ∘ is strictly increasing.* Since ∘ is continuous and cancellative, it is *strictly monotonic* in each of its two variables (factors). First we show that $x \circ y$ *cannot be increasing in* x *for one* $y = y_1$ *and decreasing for another* $y = y_2$. Indeed if $x_1 < x_2$ and

$$x_1 \circ y_1 - x_2 \circ y_1 < 0 \text{ but } x_1 \circ y_2 - x_2 \circ y_2 > 0$$

then, by the continuity of ∘ (and of subtraction) there exists an y_0 (between y_1 and y_2) such that $x_1 \circ y_0 - x_2 \circ y_0 = 0$ that is, $x_1 \circ y_0 = x_2 \circ y_0$ so, by the cancellativity, $x_1 = x_2$ contrary to $x_1 < x_2$. (*A similar statement* and proof *applies to the second variable*).

Now we can show that ∘ *cannot be decreasing* in either variable (which concludes the proof of (o)). Indeed, if for $x_1 < x_2$ we had $x_1 \circ y > x_2 \circ y$ (for all y) then we would also have $(x_1 \circ y) \circ z < (x_2 \circ y) \circ z$ (since then also $t \circ z$ would have to strictly decrease in t). But, by the associativity, this would mean $x_1 \circ (y \circ z) < x_2 \circ (y \circ z)$ so ∘ would not decrease in the first variable after all (by the first part of the proof it is not possible that $x \circ s$ would strictly decrease in x for $s = y$ but not for $s = z$ or for $s = y \circ z$). The proof that ∘ strictly decreases in the second variable is completely similar.

(i) *Classification of elements of* I. For each element c of I exactly one of the following three relations hold

$$c \circ c > c , \tag{5}$$

$$c \circ c = c , \tag{6}$$

$$c \circ c < c . \tag{7}$$

We claim that, accordingly,

$$c \circ x > x, \quad x \circ c > x \tag{8}$$

$$c \circ x = x = x \circ c \tag{9}$$

or

$$c \circ x < x, \quad x \circ c < x , \tag{10}$$

respectively, for *all* $x \in I$. Since all of these proofs are similar, we prove only that *the first inequality of (8) follows from (5)*: By (2), and since $\circ$ is increasing in the first factor,

$$c \circ (c \circ x) = (c \circ c) \circ x > c \circ x ,$$

so indeed (since $\circ$ is also increasing in the second factor)

$$c \circ x > x ,$$

as asserted. Conversely, *if*, for instance,

$$c \circ x_0 > x_0$$

holds for one $x_0 \in I$, *then*

$$(c \circ c) \circ x_0 = c \circ (c \circ x_0) > c \circ x_0$$

so $c \circ c > c$ and by what we proved above,

$$c \circ x > x \quad and \quad x \circ c > x \quad for \ all \quad x \in I ,$$

and similar statements apply to the rest of the inequalities and equations in (8), (9) or (10).

In particular, if there exists a $c \in I$ satisfying (6), it has to be the neutral element e of $(I, \circ)$ (see (9)). There is at most one neutral element in a semigroup (if there were another, e', then $e' = e' \circ e = e$). Whether there is a neutral element in I or not, since I consists of more than one point, there has to be also an element c satisfying (5) or (7). Suppose, for instance, that there exists a c satisfying (5). This will lead to the last three of the intervals J in (4) and to I

which are open from above. (The remaining cases follow if there exists a c satisfying (7). It is of course possible that there exist elements of I satisfying (5) and others satisfying (7). This would lead to $\mathbb{R}$, the last interval in (4).) Indeed,

$$c \circ c > c, \quad (c \circ c) \circ c > c \circ c, ...,$$

so

$$n \mapsto c^n \quad increases \ .$$

Here c^n (is not the ordinary power but) is defined by $c^1 = c$, $c^{n+1} = c^n \circ c$ $(n=1,2,...)$. While $c^n \in I$ for all n, we cannot have $b' = \lim_{n \to \infty} c^n \in I$ since then, by (8),

$$b' = \lim_{n \to \infty} c^{n+1}$$

$$= \lim_{n \to \infty} (c^n \circ c) = b' \circ c > b'$$

which is impossible. So

$$\lim_{n \to \infty} c^n = b \tag{11}$$

the right endpoint of I and $b \notin I$, so I *is open from above* (just as in the proof of Theorem 6.3).

(ii) *Classification according to the left endpoint of I.* We now distinguish three cases according to how its left endpoint a 'behaves'.

Case 1: The interval I is closed from the left (from below). Then its left endpoint a is in I, $a \in I$. Since $a \circ a \in I$, we must have either

$$a \circ a > a \tag{12}$$

or

$$a \circ a = a \ . \tag{13}$$

Even if the interval is open from below, the limit

$$\lim_{x \to a} (x \circ y)$$

exists (even though it is not in I; it may even be $-\infty$), since $x \circ y$ decreases as x decreases. The value of this limit distinguishes the remaining two cases.

Case 2: We have for one $y \in I$ and therefore (cf. (i)) for all $y \in I$

$$\lim_{x \to a} (x \circ y) \geq y \ \text{ so } \ \lim_{x \to a} (x \circ y) \in I \ . \tag{14}$$

(Strictly speaking we should write $\lim\limits_{x \to a+}$ but we are interested only in I and its endpoints.) In this case we get also

$$\lim_{y \to a} (x \circ y) \geq x \ \ (\text{so } \lim_{y \to a} (x \circ y) \in I) \ . \tag{15}$$

For, if we had an $x_0 \in I$ such that

$$\lim_{y \to a} (x_0 \circ y) < x_0$$

then there would be an $r \in I$ such that

$$x_0 \circ r < x_0$$

thus, by (i),

$$r \circ r < r \quad ,$$

$$r \circ y < y \quad ,$$

and so

$$\lim_{x \to a} (x \circ y) < r \circ y < y$$

contrary to (14). From either (14) or (15), we get

$$\lim_{x \to a} \lim_{y \to a} (x \circ y) \geq a \quad .$$

We show that in this case *we can join* a *to* I *without any assumption being breached.* For this purpose we define

$$a \circ t = \lim_{x \to a} (x \circ t) \in I,$$

$$t \circ a = \lim_{y \to a} (t \circ y) \in I,$$

$$a \circ a = \lim_{x \to a} \lim_{y \to a} (x \circ y) \quad .$$

By the above, $a \circ t$, $t \circ a$ $(t \in I)$ and $a \circ a$ are all in the newly extended interval $I \cup \{a\}$. It is evident that the extended operation is *continuous and associative.* We prove that it is also strictly *increasing*: If $a < s < t$ then

$$z \circ (a \circ s) = (z \circ a) \circ s < (z \circ a) \circ t = z \circ (a \circ t)$$

(because $z \circ a$ is in I and the extended operation is associative) and so, since $\circ$ is strictly increasing, $a \circ s < a \circ t$. One proves similarly $s \circ a < t \circ a$, while $a \circ a < t \circ a$, $a \circ a < a \circ t$ $(t > a)$ is obvious ($a \circ a = \lim_{t \to a}(a \circ t)$ and $t \mapsto a \circ t$ is strictly decreasing).

Case 3: The interval I is still open from below but

$$\lim_{x \to a}(x \circ y) < y \ . \tag{16}$$

In this case a plays a similar role to that of 0 in multiplication ($0.t = t.0 = 0.0 = 0$). Indeed, from (16)

$$\lim_{y \to a}\lim_{x \to a}(x \circ y) \le \lim_{y \to a} y = a$$

but $x \circ y \in I$, so its limit is at least at the left endpoint a of I, thus we have the analogue of $0.0 = 0$:

$$\lim_{x \to a}\lim_{y \to a}(x \circ y) = a \tag{17}$$

[$\circ$ being continuous, the double limit can be interchanged or replaced by

$$\lim_{\substack{x \to a \\ y \to a}}(x \circ y)] \ .$$

As to the analogue of $0.t = 0$, if we had $a' = \lim_{x \to a}(x \circ y) > a$

(it cannot be $< a$, since $x \circ y \in I$), then $a' \in I$ and, by (16),

$$a' > \lim_{s \to a}(s \circ a') = \lim_{s \to a}[s \circ \lim_{x \to a}(x \circ y)]$$

$$= \lim_{s \to a}\lim_{x \to a}[s \circ (x \circ y)]$$

$$= \lim_{\substack{s \to a \\ x \to a}}[(s \circ x) \circ y] = \lim_{t \to a}(t \circ y) = a'$$

(since, by (17), $t = s \circ x \to a$ if $s \to a$ and $x \to a$), which is impossible. So indeed

$$\lim_{x \to a}(x \circ y) = a \tag{18}$$

and, similarly,

$$\lim_{y \to a}(x \circ y) = a \quad .$$

(iii) *Reclassification, neutral and inverse elements.* We now renumber our cases. We leave case 3 alone, but in case 2 *we consider a added to I*, so it joins case 1. This united case we split again, according whether

case 1': $a \circ a > a$

or

case 2': $a \circ a = a.$

In case 1' there is no neutral element in I. Indeed, as we have seen in (i), $a \circ a > a$ implies $a \circ y > y$ for all $y > a$. Moreover $x \geq a$ for all $x \in I$ so, since $\circ$ is increasing

$$x \circ y \geq a \circ y > y \quad \text{and, similarly,} \quad x \circ y > x \quad .$$

In case 2' a is the (unique) *neutral element* as seen in (i); we will write $a = e$. But there exists no inverse element for any $y \neq e$ in I. Indeed for $y > a = e$ and any $x \in I$ (so $x \geq a = e$),

$$x \circ y \geq e \circ y = y > a = e \quad .$$

In case 3 there always exist neutral and inverse elements. If $y \circ y > y$ then, since from (16) $\lim_{x \to a}(x \circ y) < y$, there exists an e between a and y such that $e \circ y = y$. If $y \circ y < y$ then, since we have supposed in (i) the existence of a c with $c \circ c > c$, $c \circ y > y$, there exists an e between y and c such that $e \circ y = y$. The e is unique and also satisfies $e \circ t = t \circ e = t$ for all $t \in I$, as we have seen in (i). Also $e > a$ ($a \notin I$ in this case 3). We have, for an arbitrary $y > e$, $e \circ y > e$. But, by (18), $\lim_{x \to a}(x \circ y) = a < e$. So between a and e there exists a y^{-1} such that

$$y^{-1} \circ y = e; \text{ then also } y \circ y^{-1} = e \quad.$$

Similarly, there also exists such an inverse element for $x < e$ because $x \circ e < e$ and, with the c satisfying $z \circ c > z$ for all $z \in I$, whose existence we have supposed in (i), $\lim_{n \to \infty} (x \circ c^n) = b > e$. Indeed $x \circ c^{n+1} = (x \circ c^n) \circ c > x \circ c^n$, so $n \mapsto x \circ c^n$ increases and $b'' = \lim_{n \to \infty} (x \circ c^n)$ exists. If we had $b'' \in I$ then $b'' = \lim_{n \to \infty} (x \circ c^{n+1}) = \lim_{n \to \infty} (x \circ c^n) \circ c > b'' \circ c > b''$ which is impossible, so $\lim_{n \to \infty} (x \circ c^n) = b > e$. So there exists an $x^{-1} > e$ such that $x \circ x^{-1} = e$ (and then, again, $x^{-1} \circ x = e$ too). Of course, for all $y > e$ we have $y \circ e > e$ so $y \circ y > y$ by (i) and, similarly, for all $x < e$, $x \circ e < e$ and $x \circ x < x$. If $c > e$, then $c^{-1} < e$ and, similarly to (11), we have

$$\lim_{n \to \infty} (c^{-1})^n = a \qquad (19)$$

and *I is open from below.*

(iv) *'Powers' and 'roots'.* Examining further the 'power functions' defined, in accordance with (i) as

$$x^1 = x, \ x^{n+1} = x^n \circ x \quad (n=1,2,...) \ , \qquad (20)$$

we notice that they satisfy

$$x^m \circ x^n = x^{m+n} = x^n \circ x^m \qquad (21)$$

and

$$(x^m)^n = x^{mn} = (x^n)^m \quad (m,n \in I\!N) \qquad (22)$$

and also that the notation x^{-1} for the inverse element of x does not conflict with this notation. Quite the contrary: if we define

$$x^{-n} = (x^{-1})^n \quad (n \in I\!N), \quad x^\circ = e \tag{23}$$

then (21) and (22) are satisfied for all integers m,n. We want to go over to rational exponents through the 'root functions', the inverse functions of the powers. Since $\circ$ is continuous and strictly increasing, so are $x \mapsto x^n$ $(n=1,2,...)$. The function $x \mapsto x^n$ *maps I in the case 1' onto* $[a^n,b[$, *in the case 2' onto* $]a,b[$, *and in the case 3 onto* $]a,b[$ *(because of (11) and (19)). So* $y \mapsto {}^n\!\sqrt{y}$ *is uniquely defined, continuous and strictly increasing on* $[a^n,b[$, $[a,b[$ *or* $]a,b[$, *respectively. Also* ${}^n\!\sqrt{c} > e$ *if* $c > e$.

(v) *Construction of* ϕ *on the rationals.* We have arrived now to the point where we start to construct the function ϕ for (3). We define $\phi(1) = c$ with a c for which $c \circ c > c$ (by our assumptions such a $c \in I$ exists), with no further restrictions in cases 2' and 3, but specifically we choose $c = a$ in the case 1'.

Then we define

$$\phi(\frac{m}{n}) = ({}^n\!\sqrt{c})^m$$

$$(m,n \in I\!N) \text{ and } \phi(0) = e \tag{24}$$

in cases 2' and 3 (this definition is unambiguous because of (22)). In the case 3 we also define

$$\phi(-\frac{m}{n}) = \phi(\frac{m}{n})^{-1} = ({}^n\!\sqrt{c})^{-m} \tag{25}$$

[cf. (23)]. One sees that this ϕ, defined in case 2' for the non-negative rationals, in case 3 for all rationals, is strictly increasing:

$$\phi(\frac{m+1}{n}) = (^n\sqrt{c})^{m+1} > (^n\sqrt{c})^m$$

$$= \phi(\frac{m}{n}) > e = \phi(0) > \phi(\frac{m}{n})^{-1}$$

$$= \phi(-\frac{m}{n}) > \phi(-\frac{m+1}{n})$$

and (cf. (11))

$$\lim_{r \to \infty} \phi(r) = b \quad (r \in \mathbb{Q}) \tag{26}$$

in both cases, and in case 3 also (cf. (19))

$$\lim_{r \to -\infty} \phi(r) = a \quad (r \in \mathbb{Q}) \quad . \tag{27}$$

Furthermore, (24) and (21) imply

$$\phi(r+s) = \phi(r) \circ \phi(s) \tag{28}$$

for all nonnegative rationals in case 2' and all $r,s \in \mathbb{Q}$ in case 3.

In case 1' things are a bit more complicated but the results are the same. We give here only a sketch and refer the reader to Aczél [66]. Here $c = a$ and so $^n\sqrt{c}$ is not defined since the n-th root function is defined only on $[a^n,b[$. So we replace (24) by

$$\phi(\frac{m}{n}) = {}^n\sqrt{(c^m)} \quad \text{for} \quad m \geq n \quad (m,n \in \mathbb{N}) \quad .$$

We still have (26), don't need (27), ϕ still strictly increases on the rationals, however (28) (for $r,s \in \mathbb{Q} \cap [1,\infty[$ is more difficult to prove. Mainly the lack of a commutativity supposition is disturbing. So one proves in succession

$$\phi(m) \circ \phi(\frac{m}{n}) = \phi(\frac{(n+1)m}{n}) = \phi(\frac{m}{n}) \circ \phi(m) \quad ,$$

$$\phi(\frac{m_1}{n}+\frac{m_2}{n})\circ\phi(m_1)\circ\phi(m_2) \; = \; \phi(m_1)\circ\phi(\frac{m_1}{n})\circ\phi(\frac{m_2}{n})\circ\phi(m_2)$$

$$= \; \phi(m_1)\circ\phi(m_2)\circ\phi(\frac{m_1+m_2}{n}) \quad ,$$

$$\phi(\frac{m_1}{n}+\frac{m_2}{n})\circ\phi(m_1) \; = \; \phi(\frac{m_1}{n})\circ\phi(m_1)\circ\phi(\frac{m_2}{n}) \quad ,$$

$$\phi(\frac{m}{n})\circ\phi(p) \; = \; \phi(\frac{m}{n}+p) \; = \; \phi(p)\circ\phi(\frac{m}{n}) \quad ,$$

$(m,m_1,m_2,n,p$ are positive integers) and the last two equations already give

$$\phi(\frac{m_1+m_2}{n}) \; = \; \phi(\frac{m_1}{n})\circ\phi(\frac{m_2}{n}), \; (m_1 \geq n, \, m_2 \geq n; \, m_1,m_2,n \in I\!\!N) \quad ,$$

that is (28) for all $r,s \in \mathbb{Q} \cap [1,\infty[$.

(vi) *Extension of ϕ to the reals.* We have now (28) for all rationals not smaller than 1 in case 1'. for all nonnegative rationals in case 2' and for all rationals in case 3. We have seen also that ϕ is strictly increasing for these rationals. We now extend the definition of ϕ to all reals in $[1,\infty[$, to all of $I\!\!R_+$ or to all $I\!\!R$, respectively. Let u be an arbitrary real number in the interior of the appropriate intervals (we know already the values of ϕ on their closed ends). There exist an increasing and a strictly decreasing sequence of rationals $\{r_n\}$ and $\{R_n\}$ with

$$\lim_{n \to \infty} r_n \; = \; u \; = \; \lim_{n \to \infty} R_n \quad .$$

So

$$r_1 \leq r_2 \leq ... \leq r_n \leq r_{n+1} \leq ... \leq u$$

$$<...< R_{n+1} < R_n <...< R_2 < R_1 \quad .$$

Since ϕ is increasing on these rationals,

$$\phi(r_1) \leq ... \leq \phi(r_n) \leq \phi(r_{n+1})$$
$$\leq ... \leq \phi(R_{n+1}) \leq \phi(R_n) \leq ... \leq \phi(R_1) \ .$$

So $\{\phi(r_n)\}$ is increasing and bounded from above by $\phi(R_m)$, while $\{\phi(R_n)\}$ is decreasing and bounded from below by $\phi(r_m)$ ($m \in I\!N$ arbitrary). Therefore both are convergent and

$$\lambda = \lim_{n \to \infty} \phi(r_n) \leq L = \lim_{n \to \infty} \phi(R_n) \ . \tag{29}$$

Since f is increasing, *the limits will be the same λ and L for sequences of other rationals tending to u from below or from above, respectively.*

We prove that $\lambda = L$. Indeed choose sequences of rationals $\{r_n'\}$ and $\{R_n'\}$, tending to u increasingly or decreasingly, respectively, and such that

$$\rho_n = \frac{r_n' + R_n'}{2} > u \ . \tag{30}$$

By (28)

$$\phi(r_n') \circ \phi(R_n') = \phi(2\rho_n) = \phi(\rho_n) \circ \phi(\rho_n) \ .$$

In view of (29) and (30) this goes over, as $n \to \infty$, into

$$\lambda \circ L = L \circ L$$

and, since $\circ$ is strictly increasing (and cancellative), we get

$$\lambda = L$$

as asserted. We assign this $\lambda = L$ as the value of ϕ at u.

The function ϕ, thus extended, is still strictly increasing. Indeed, take arbitrary $u < v$ (in $[1, \infty[$, $I\!\!R_+$ or $I\!\!R$, as may be the case). Let R be an upper rational approximation of u (or just a rational number > 1 or > 0 if $u = 1$ or $u = 0$ in the cases 1' and 2', respectively) and r a lower rational approximation of v such that $R < r$. Then

$$\phi(u) \leq \phi(R) < \phi(r) \leq \phi(v)$$

as asserted (since ϕ is strictly increasing for the rationals R,r).

As increasing function, ϕ can have at most gap singularities. If it had one, there $\lambda < L$ (cf. (29)) would hold. But we just proved that $\lambda = L$. So ϕ is *continuous* (the proof of the continuity from the right at 1 or 0 in the cases 1' and 2', respectively, is also easy). Therefore (28) extends to

$$\phi(u+v) = \phi(u) \circ \phi(v) \tag{31}$$

for all real u,v (≥ 1, ≥ 0 or arbitrary). We have chosen $\phi(1) = a$ in case 1', $\phi(0) = a$ in case 2' and proved (27) $\lim\limits_{r \to -\infty} \phi(r) = a$ in case 3. In all cases we had

$$\lim_{r \to \infty} \phi(r) = b \quad .$$

So, for arbitrary $x,y \in I$ there exist u,v in the appropriate interval such that $x = \phi(u)$, $y = \phi(v)$. Thus we indeed get, as asserted in the theorem,

$$x \circ y = \phi[\phi^{-1}(x) + \phi^{-1}(y)] \text{ for all } x,y \in I \quad . \tag{3}$$

(vii) *The domains of* ϕ. The domains of ϕ were the intervals $[1,\infty[$, $[0,\infty[$ and $]-\infty,\infty[= \mathbb{R}$ in the cases 1', 2' and 3, respectively. Taking our original case 2 into consideration, we get also $]1,\infty[$ and $]0,\infty[$ and, finally, if there was no $c \in I$ satisfying (5) but there was one satisfying (7), we would have got also $]-\infty,-1]$, $]-\infty,-1[$, $]-\infty,0[$ and $]-\infty,0[$ as domains. Let us call $\tilde{J}$ any of these intervals.

(viii) *Uniqueness*. Finally, we examine the uniqueness

of ϕ: We use the inverse $\psi: I \to \tilde{J}$ of ϕ which exists as we have seen and is also continuous and strictly increasing. So we write (3) in the form

$$x \circ y = \psi^{-1}[\psi(x)+\psi(y)] \quad (x,y \in I) \quad .$$

If there existed another continuous and strictly monotonic (but not necessarily increasing) function $\widetilde{\psi}$ such that

$$\psi^{-1}[\psi(x)+\psi(y)] = \widetilde{\psi}^{-1}[\widetilde{\psi}(x)+\widetilde{\psi}(y)] \quad (x,y \in I) \qquad (32)$$

then we write $u = \psi(x)$, $v = \psi(y)$ and get

$$\widetilde{\psi}[\psi^{-1}(u+v)] = \widetilde{\psi}[\psi^{-1}(u)] + \widetilde{\psi}[\psi^{-1}(v)] \quad \text{on} \quad \tilde{J}^2 \quad .$$

($\tilde{J}$ being one of the intervals defined in (vii)). This is a restricted Cauchy equation with $\tilde{J}^2$ as domain. As we have seen, $\tilde{J}$ goes into infinity at least on one side, so $(\tilde{J}^2)_x \cap (\tilde{J}^2)_y \cap (\tilde{J}^2)_{x+y}$, as defined in (5.16), is certainly nonempty. Also, $\tilde{J}^2$ is a region with possibly two borderlines attached (if $\tilde{J}$ is closed on the other side). If we leave these borderlines aside for a moment, then a combination of Corollaries 5.6, 1.5 and of the remark after the former gives that

$$\widetilde{\psi}[\psi^{-1}(u)] = Cu \qquad (33)$$

($C \neq 0$ since $\widetilde{\psi}[\psi^{-1}(u)]$ is strictly monotonic) .

Since both ψ and $\widetilde{\psi}$ are continuous, this remains true for the whole of $\tilde{J}$ (including the endpoint possibly omitted before), that is,

$$\widetilde{\psi}(x) = C\psi(x) \quad \text{for all} \quad x \in I, (C \neq 0) \qquad (34)$$

(which, of course, *always satisfies (32)*) or, going over to inverses,

$$\tilde{\phi}(u) = \phi(Cu) \text{ for all } u \in \tilde{J} \ (C \neq 0) \ ,$$

as asserted in the last sentence of Theorem 1. With the transformation $u \mapsto Cu \ (C \neq 0)$, the intervals $\tilde{J}$ in (vii) go over into the intervals J as defined in (4).

The theorem is proved.   □

Remark. In (ii) case 2 we can join the left endpoint a (or in the dual case, the right endpoint b) to I even if $a = -\infty$ ($b = \infty$), which is possible since we allowed infinite intervals. Then $a = -\infty$ will behave as any finite element of I. For instance, for

$$x \circ y = \log(e^x + e^y) \ ,$$

$x = -\infty$ may be considered as a quite ordinary (in this case neutral) element of I.

We apply this result to the synthesis of judgements in the following sense (Saaty [80]).

We have n quantifiable judgements $x_1,...,x_n$ $(n \geq 2)$ which we want to synthesize into a 'consensus' judgement $f(x_1,...,x_n)$. Differing from section 1 we have no fixed amount to distribute, the judgements are all about the same object or, more typically, *ratio*-judgements comparing two objects (how many times heavier or, more imponderably, how many times better, more important etc. one object or alternative is than the other). It is supposed that $x_k \in P$ $(k=1,2,...,n)$ where P is a proper real interval. It is also supposed that f is *separable*

$$f(x_1,x_2,...,x_n) = g(x_1) \circ g(x_2) \circ \cdots \circ g(x_n)$$

$$(x_1,x_2,...,x_n \in P) \ . \tag{35}$$

Here g is a *continuous nonconstant* function so the image of P under g, $I = g(P)$ is itself a proper interval. It is further supposed that $\circ$ is a *continuous, associative and cancellative* operation. In Theorem 1 we have proved that *it follows from cancellativity, continuity and associativity* that $\circ$ is *strictly increasing* in each 'factor'. (The problem was solved under weaker conditions in Aczél [84], permitting even different g's for each variable in (35).) - Since $\circ$ is associative, the right hand side of (35) makes perfectly good sense without further brackets.

As we have seen in Theorem 1 (see (3)), our conditions concerning $\circ$ imply that

$$y \circ z = \psi^{-1}[\psi(y) + \psi(z)] \quad (y, z \in I)$$

where $\psi : I \to J$ is a continuous strictly monotonic function, the interval I has to be *open or half-open* and *J is one of the intervals (4)*. So (35) goes over into

$$f(x_1, ..., x_n) = \psi^{-1}\left(\sum_{k=1}^{n} \psi[g(x_k)]\right)$$

$$(x_1, ..., x_n \in P) . \tag{36}$$

Since $I = g(P)$ and g is continuous and strictly increasing with I also P is an open or half-open interval.

Our next, *unanimity* (or 'consensus', cf. (1.3), (1.4)) condition

$$f(x, ..., x) = x \quad (x \in P) \tag{37}$$

is also quite natural: if all individual judgements have the same value x then the value of the synthesized judgement should be x too. Combined with (36) we get

$$\psi(x) = n\psi[g(x)] ,$$

that is,

$$g(x) = \psi^{-1}(\frac{\psi(x)}{n}) \ (x \in P) \ . \tag{38}$$

(Notice that (37) connects P with the range of f which is the domain of ψ. So ψ is now considered on P.) So also g *is strictly increasing* and

$$f(x_1,...,x_n) = \psi^{-1}(\frac{1}{n} \sum_{k=1}^{n} \psi(x_k))$$

$$(x_k \in P; \ k=1,...,n) \ . \tag{39}$$

Functions of the form (39) are called *quasiarithmetic means* (for obvious reasons, since $\dfrac{1}{n} \sum\limits_{k=1}^{n} y_k$ is the arithmetic mean of $y_1,...,y_n$).

If, as mentioned above, we deal with ratio-judgements, then one supposes also the *reciprocal* condition

$$f(1/x_1,...,1/x_n) = 1/f(x_1,...,x_n)$$

$$(x_k \in P \Rightarrow 1/x_k \in P; \ k=1,...,n) \ . \tag{40}$$

For instance, if the individual judgements are about how many times heavier the object A is than the object B, then their values change into reciprocals if B is compared to A. It is reasonable to suppose that also the synthesized judgement changes into its reciprocal and this is expressed by (40). Note that with x_k also $1/x_k$ has to be in the interval P. Therefore P cannot be half-open, it contains 1, does not contain 0, so consists of positive numbers (which is again quite natural). So $P \subseteq \mathbb{R}_{++}$ is an *open* interval which contains with each element also its reciprocal. Consequently also $I = g(P)$ *and* $J = \psi(I)$ *are open* (since g and ψ are continuous and strictly monotonic) and so the possibilities for J are reduced to the open intervals in (4), that is

$$J = \,]-\infty,\gamma[\quad \text{or} \quad]\delta,\infty[\quad \text{or} \quad]-\infty,\infty[\,= \mathbb{R} \quad . \tag{41}$$

Let us put now (39) into (40)

$$\psi^{-1}[\frac{1}{n}\sum_{k=1}^{n}\psi(\frac{1}{x_k})] = 1/\psi^{-1}[\frac{1}{n}\sum_{k=1}^{n}\psi(x_k)]$$

$$(x_k \in P; \; k=1,...,n) \quad . \tag{42}$$

In order to solve this equation we need a result on the uniqueness of ψ in (39), similar to (but also somewhat different from) the uniqueness statement in Theorem 1.

Lemma 2. *Let P be an interval and $\psi:P \to J$, $\widetilde{\psi}:P \to \tilde{J}$ be continuous and strictly monotonic functions. Then*

$$\psi^{-1}[\frac{1}{n}\sum_{k=1}^{n}\psi(x_k)] = \widetilde{\psi}^{-1}[\frac{1}{n}\sum_{k=1}^{n}\widetilde{\psi}(x_k)] \tag{43}$$

if, and only if, there exist constants $\alpha \neq 0$ and β such that

$$\widetilde{\psi}(x) = \alpha\psi(x) + \beta \quad (x \in P) \quad . \tag{44}$$

Proof. The "if" part is obvious. In order to prove the "only if" part, we put into (43)

$$y_k = \psi(x_k) \in J \quad (k=1,...,n)$$

and define Ψ by

$$\Psi(y) = \widetilde{\psi}[\psi^{-1}(y)] \quad (y \in J) \quad , \tag{45}$$

in order to get

$$\Psi(\frac{1}{n}\sum_{k=1}^{n}y_k) = \frac{1}{n}\sum_{k=1}^{n}\Psi(y_k)$$

$$(y_k \in J;\ k=1,...,n)\ .\qquad\qquad(46)$$

This is the *n-term Jensen equation* (the 2-term Jensen equation is called simply the *Jensen equation*). By putting into it $y_3 =...= y_n = c$ (constant) and writing

$$\Phi(y_2) = \Psi(y_2) + (n-2)\Psi(c)\ ,$$

$$h(y) = n\Psi(\frac{1}{n}[y+(n-2)c])\ ,$$

(46) is transformed into

$$h(y_1+y_2) = \Psi(y_1) + \Phi(y_2)\ ((y_1,y_2) \in J^2)\ .\qquad(47)$$

This is a *restricted Pexider equation*. If, as in our case, J is open, then J^2 is a region and, by Theorem 5.4

$$\Psi(y) = \phi(y) + \beta\ (y \in J)\ ,\qquad\qquad(48)$$

where β is an arbitrary constant and ϕ an arbitrary solution of Cauchy's equation

$$\phi(y_1+y_2) = \phi(y_1) + \phi(y_2)\ (y_1,y_2 \in \mathbb{R})\ .\qquad(49)$$

By Corollary 1.5, if Ψ and thus ϕ is continuous at a point (which is so in our case, even everywhere), then there exists a constant α such that

$$\Psi(y) = \alpha y + \beta\ (y \in J)\ .\qquad\qquad(50)$$

(*If Ψ is continuous on J then the interval J could be also half-open or closed*, since (50) can then be extended to the endpoints of J).

In view of the definition (45) of Ψ, we have indeed

$$\widetilde{\psi}(x) = \alpha\psi(x) + \beta \quad (x \in P)$$

and, since $\widetilde{\psi}$ is strictly monotonic, $\alpha \neq 0$.   □

Corollary 3. *The general solution of the Jensen equation (46) for any fixed $n \geq 2$ and for any open real interval J is given by (48) and its general solution continuous at a point (or bounded from one side on a proper interval etc.) by (50). Here α, β are arbitrary real constants and ϕ an arbitrary solution of the Cauchy equation (49). The general solution, continuous at the endpoint(s) of J belonging to J are given by (50) also if J is half-closed or closed.*

Returning to our equation (42), if we write there $\widetilde{\psi}(x) = \psi(\frac{1}{x})$, then it goes over into (43). So by Lemma 2 we have

$$\psi(\frac{1}{x}) = \alpha\psi(x) + \beta \quad (x \in P) \ . \tag{51}$$

Repeated use of this equation gives

$$\psi(x) = \alpha\psi(\frac{1}{x}) + \beta = \alpha^2\psi(x) + \alpha\beta + \beta^2 \ .$$

Since ψ is not constant, this implies

$$\alpha^2 = 1, \quad \alpha\beta + \beta = 0 \ ,$$

that is, either $\alpha = 1$, $\beta = 0$ or $\alpha = -1$ (in which case $\alpha\beta + \beta = 0$ is satisfied for every β). The first case would change (51) into $\psi(1/x) = \psi(x)$ $(x \in J)$ and ψ would not be strictly monotonic. So

$$\psi(\frac{1}{x}) = -\psi(x) + \beta \ . \tag{52}$$

Introducing $t = \log x$ (remember that P consisted of positive numbers) and the function ω by

$$\omega(t) = \psi(e^t) - \frac{\beta}{2}$$

$$(t \in P' = \log P = \{\log u \mid u \in P\}) \ , \tag{53}$$

equation (52) goes over into

$$\omega(-t) = -\omega(t) \quad (t \in P') \ ,$$

that is, ω is *odd* (since P contains with every element its reciprocal, so P' *is symmetric with respect to* 0). By (53), $\psi(x) = \omega(\log x) + \dfrac{\beta}{2}$ and by (39) we have

$$f(x_1,...,x_n) = \exp \omega^{-1}[\frac{1}{n} \sum_{k=1}^{n} \omega(\log x_k)]$$

$$(x_k \in P; \ k=1,...,n) \ . \tag{54}$$

Conversely, (54) with an arbitrary continuous, strictly monotonic and odd ω satisfies (35), (37) and (40). So we have proved the following.

Theorem 4. *Let f be of the form (35) where P is a proper real interval, $g: P \rightarrow I$ $(I = g(P))$ is continuous, nonconstant and $\circ: I \times I \rightarrow I$ is continuous, cancellative and associative. The unanimity and reciprocal conditions (37) and (40) are satisfied if, and only if, f is of the form (54) where ω is an arbitrary continuous and strictly monotonic odd function. — Under these conditions P has to be open, $P \subseteq \mathbb{R}_{++}$ (and $x \in P \Rightarrow 1/x \in P$, so also $1 \in P$).*

(For similar but somewhat more general results see

Aczél-Saaty [83], Aczél [84].)

Saaty [80] has actually used geometric means for synthesizing judgements. This is evidently the special case $\omega(x) = x$ of (54). We will give now a further condition which narrows down the solution to this special case. This condition will be the first degree *homogeneity* which has already played a crucial role in sections 5 and 6 (and minor roles also in sections 2 and 3). For ratio judgements it means that, if each individual judges a second ratio r times as large as the first ratio, then also the synthesized judgement on the value of the second should be r times as large as that on the first. So we have

$$f(rx_1, rx_2, ..., rx_n) = rf(x_1, x_2, ..., x_n)$$

$$(r > 0,\ x_k, rx_k \in P) \ . \tag{55}$$

As we see, this is a *restricted homogeneity* equation, since rx_k has to remain in P. (We would have arrived at similar conditional homogeneity conditions in sections 3, 5 and 6 had we not allowed the variables to roam over all of $\mathbb{R}_{++}$.)

We will examine the consequences of (55) for (39), rather than just for (54), that is. *we will determine all quasiarithmetic means homogeneous of first degree* on an interval $P \subseteq \mathbb{R}_{++}$ which contains 1. (For the case of $x_1, ..., x_n \in \mathbb{R}_{++}$ in (39) and (55) see for instance Hardy-Littlewood-Pólya [52].) Here we will not suppose that the interval $P \subseteq \mathbb{R}_{++}$ is open. So we have

$$\psi^{-1}[\frac{1}{n} \sum_{k=1}^{n} \psi(rx_k)] = r\psi^{-1}[\sum_{k=1}^{n} \psi(x_k)]$$

$$(r > 0;\ x_k,\ rx_k \in P;\ k = 1, ..., n) \ , \tag{56}$$

where P is any proper interval of positive numbers. If we hold u temporarily constant, we see that this again is an equation of the form (43) with $\widetilde{\psi}(x) = \psi(rx)$. By Lemma 2 we get

$$\widetilde{\psi}(x) = \alpha\psi(x) + \beta \quad (\alpha \neq 0)$$

or, taking the dependence upon r into consideration again

$$\psi(rx) = \alpha(r)\psi(x) + \beta(r) \ . \tag{57}$$

Strictly speaking, (56) is of the form (43) only for $x_k \in P \cap \dfrac{1}{r}P$, where $\dfrac{1}{r}P = \{x \mid rx \in P\}$. So we get (57) exactly for all $r > 0$, $x \in P$, such that also $rx \in P$. We will even restrict (57) a bit further and consider

$$\psi(rx) = \alpha(r)\psi(x) + \beta(r)$$

$$\text{whenever } x,r,xr \in P \ . \tag{58}$$

Equation (58) is a *restricted linear-affine equation*, while (2.5) was the unrestricted one. Remembering that we have $1 \in P$, we treat this equation as in section 2 with the following changes (cf. Stehling [74, 75], Aczél [84]). In place of equation (2.8) we have now for $\overline{\psi}(x) \equiv \psi(x) - \psi(1)$ the *restricted logarithmic equation*

$$\overline{\psi}(rx) = \overline{\psi}(x) + \overline{\psi}(r) \quad (x,r,xr \in P) \ ,$$

which we reduce, as there, with $r = e^s$, $x = e^t$ to the restricted Cauchy equation (1.38) and by Corollary 1.8 we get, since ψ is strictly monotonic, $\overline{\psi}(x) = A\log x$ and

$$\psi(x) = A\log x + B \text{ for all } x \in P \ (A \neq 0) \ . \tag{59}$$

Then again everything goes as in section 2 except that (2.13) is replaced by the *restricted multiplicative equation*

$$\alpha(rx) = \alpha(r)\alpha(x) \quad (x, r, xr \in P) \ . \tag{60}$$

In order to reduce this to a restricted Cauchy equation, we have only to see that we can take logarithms on both sides again. Since $1 \in P$, whenever $x \in P$, also $\sqrt{x} \in P$, so we have again

$$\alpha(x) = \alpha(\sqrt{x})^2 \geq 0 \quad \text{for all} \quad x \in P \ . \tag{61}$$

If $\alpha(1) = 0$ then by (60) $\alpha(x) = 0$ for all P contrary to the strict monotonicity of α (a consequence of that of ψ). If there is in P an $x_o \neq 1$ such that $\alpha(x_o) = 0$ then by (61) also $\alpha(\sqrt{x_o}) = 0$. Since $\sqrt{x_o} \neq x_o$ if $x_o \neq 1$, this would again contradict the strict monotonicity of α. So $\alpha(x) > 0$ on P, we can take logarithms of both sides of (60) and reduce it eventually to (1.38) and get, by Corollary 1.8

$$\alpha(x) = x^c \ , \quad \bar{\psi}(x) = \psi(x) - \psi(1) = Ax^c + b$$

and

$$\psi(x) = Ax^c + B \quad \text{for all} \quad x \in P \quad (A \neq 0, \ c \neq 0) \ . \tag{62}$$

So we have the following.

Lemma 5. *If P is an interval of positive numbers containing 1 then the general strictly monotonic solutions $\psi: P \rightarrow \mathbb{R}$ of (58) are given by (59) and (62).*

If we put (59) and (62) into (39) we get the following.

Corollary 6. *The general quasiarithmetic means (39) (with continuous strictly monotonic ψ and P an interval with $1 \in P \subseteq \mathbb{R}_{++}$), which satisfy the restricted homogeneity equations, are given by the root-mean-powers*

$$f(x_1,...,x_n) = (\frac{1}{n} \sum_{k=1}^{n} x_k^C)^{1/C}$$

$$(x_1,...,x_n \in P) \quad with \quad C \neq 0 \qquad (63)$$

and by the geometric mean

$$f(x_1,...,x_n) = (\prod_{k=1}^{n} x_k)^{1/n} \quad (x_1,...,x_n \in P) \ . \qquad (64)$$

We now go back to our synthesizing functions f. We had even without (40) that P is half-open or open. On the other hand (63) does not satisfy (40) except if $x_1 = x_2 =...= x_n$. So we have the following.

Corollary 7. *The synthesizing function f is of the form (35) (where $P \subseteq \mathbb{R}_{++}$ is an interval containing 1, g is continuous, nonconstant and $\circ$ is continuous, cancellative and associative) and satisfies the unanimity and homogeneity conditions (37) and (55) if, and only if, (63) or (64) holds. Under these conditions P has to be open or half-open. If, in addition, the reciprocal condition (40) is satisfied for at least one n-tuple $(x_1,...,x_n) \in P^n$ with not all x_k equal, then (64) has to hold for all $(x_1,...,x_n) \in P^n$.*

We note also that *in Corollary 7 the unanimity condition (37) can be restricted to*

$$f(1,...,1) = 1 \qquad (65)$$

(which is also the $(x_1,...,x_n) = (1,...,1)$ case of the reciprocal condition (40) since, by (39), $f(x_1,...,x_n) \in P \subseteq \mathbb{R}_{++}$, so $f(1,...,1) \neq -1$). Indeed, from (65) and (55) we have (37):

$$f(x,...,x) = xf(1,...,1) = x \quad for \ all \quad x \in P \ .$$

8. Synthesis of measure judgements. Equations in a single variable. The Abel and Schröder equations. Iteration.

We have seen in the previous section that for ratio judgements the reciprocal condition (7.40) is very natural, but the homogeneity condition (7.55) may be less so. If, however, the judgements to be synthesized are measures (weights, lengths, monetary values etc.) rather than their ratios, then the homogeneity condition is very natural, it expresses the property of *scale-invariance* now quite familiar to us (sections 2,3): If the scale (unit of measurement) is changed in the individual judgements, the same change of scale should occur in the synthesized judgement. On the other hand, in this case the reciprocal condition is not so natural. So for synthesizing measure judgements one could use just (7.55) and Corollary 7.6 or the first part of Corollary 7.7.

Moreover, this brings up the question of replacing the condition (7.40) by a more general one. If, quite generally, one replaces $x \mapsto x^{-1}$ in (7.40) by $h : P \longrightarrow P$, that is, we have

$$f[h(x_1),...,h(x_n)] = h[f(x_1,...,x_n)]$$

$$(x_1,...,x_n \in P) , \tag{1}$$

then (7.42) is replaced by

133

$$\psi^{-1}\Big(\frac{1}{n}\sum_{k=1}^{n}\psi[h(x_k)]\Big) = h\Big(\psi^{-1}\Big[\frac{1}{n}\sum_{k=1}^{n}\psi(x_k)\Big]\Big)$$

$$(x_k, h(x_k) \in P; \; k=1,...,n) \; . \tag{2}$$

If we write here $y_k = \psi(x_k) \in J$ (where P and thus J are, say, open intervals) and define Ψ by

$$\Psi(y) = \psi(h[\psi^{-1}(y)]) \quad (y \in J) \tag{3}$$

then we get

$$\Psi\Big(\frac{1}{n}\sum_{k=1}^{n}y_k\Big) = \frac{1}{n}\sum_{k=1}^{n}\Psi(y_k)$$

$$(y_k \in J; \; k=1,...,n) \; . \tag{4}$$

This is the Jensen equation (7.46). Now, *if h is continuous at a point* (or bounded from one side on a proper interval, etc.), so is Ψ in (3) (since ψ is continuous and strictly monotonic). Therefore, by Corollary 7.3,

$$\Psi(y) = \alpha y + \beta \; \text{ for all } \; y \in J \; .$$

From (3) we obtain with $y = \psi(x)$

$$\psi[h(x)] = \alpha\psi(x) + \beta \quad (x \in P) \; . \tag{5}$$

So *the general continuous and strictly monotonic ψ satisfying (2) for a given h, which is continuous at a point, is a solution of (5).*

The equation (5) is of a different kind than all our previous functional equations. Those, for instance the single- and multiplace Cauchy equations (1.23) and (1.19), the Jensen equation (4), the Pexider equation (5.1), the linear-affine equation (2.5), the translation equation (6.21), the associativity equation (7.1), the homogeneity equation (7.55), the generalized homogeneity (3.23), etc. contained *more than one variable*, actually *more variables than the number of places in the unknown function(s)*. (The equation (7.40) is an exception from the latter observation, but we considered it only in combination with other equations which did contain more variables than the number of places in the unknown function.) The functional equation (5) *contains only one variable*. Such equations (and by extension also those which do not contain more variables than the number of places in the unknown functions) are called *equations in a single variable*. For the study of such equations we refer the reader to Kuczma [68]. Here we mention only that (5) leads to the two functional equations in a single variable which are probably the most important, the Abel and Schröder equation. We will also give the general solution of (5) in some special cases of interest for our synthesizing problem.

We get from (5), if $\alpha = 1$, the *Abel equation*

$$\psi[h(x)] = \psi(x) + \beta \tag{6}$$

and, if $\alpha \neq 1$, the *Schröder equation*

$$\phi[h(x)] = \alpha\phi(x) \tag{7}$$

for $\phi(x) = \psi(x) + \dfrac{\beta}{\alpha-1}$. These equations are important also in *connection with the translation equations* (6.6) and (6.21) to the *theory of iteration*.

If I is an interval, I' a subinterval of I and $h : I \to I'$ a function then the *iterates* $h^1, h^2, h^3, \ldots$ of h are defined by

$$h^1(x) = h(x), \quad \text{and} \quad h^{m+1}(x) = h[h^m(x)]$$

$$(m = 1, 2, \ldots; \; x \in I) \; . \tag{8}$$

They satisfy

$$h^{m+n}(x) = h^n[h^m(x)]$$

$$(x \in I; \; m,n \in I\!N) \; . \qquad\qquad (9)$$

The problem of *continuous iteration* consists of imbedding h^n into a function $\Psi : I \times I\!R \longrightarrow I$ so that

$$\Psi(x,n) = h^n(x) \quad \text{for} \quad n \in I\!N \quad (\text{and } x \in I) \qquad (10)$$

and also (9) is extended to

$$\Psi(x,s+t) = \Psi[\Psi(x,s),t] \quad (x \in I; \; s,t \in I\!R) \; . \qquad (11)$$

But this is the *translation equation* (6.21). As we have seen in Theorem 6.3, *if* Ψ *is continuous in both variables and nonconstant in the second then there exists a continuous and strictly monotonic* $\psi : I \longrightarrow I\!R$ *such that*

$$\Psi(x,s) = \psi^{-1}[\psi(x)+s] \quad (x \in I, \; s \in I\!R) \; . \qquad (12)$$

(In (6.22) we had $\psi = \lambda^{-1}$.) Now, (10) and (11) imply (9) which in turn gives, for $n = 1$, $h^{m+1}(x) = h^1[h^m(x)]$, so that we have (8) if only $h^1(x) = h(x)$, that is, in view of (10) and (12)

$$h(x) = \Psi(x,1) = \psi^{-1}[\psi(x)+1] \; . \qquad\qquad (13)$$

But this means

$$\psi[h(x)] = \psi(x) + 1 \quad (x \in I) \; ,$$

which is an *Abel equation* (6) with $\beta = 1$. Note that, from (12), $\Psi(x,0) = x$ which conforms to the usual definition $h^{\circ}(x) = x$ and from (11)

$$x = \Psi(x,0) = \Psi[\Psi(x,1),-1] = \Psi[h(x),-1]$$

in accordance with $h^{-1}[h(x)] = x$ which shows that the inverse function h^{-1} (if it exists) can indeed be considered as (-1)-st iterate of h just as

$$x = \Psi(x,0) = \Psi[\Psi(x,n),-n] = \Psi[h^n(x),-n]$$

suggests

$$h^{-n}[h^n(x)] = x \ ,$$

that is, h^{-n} is the inverse of h^n (if it exists).

Similarly, for the multiplicative translation equation

$$G(x,uv) = G[G(x,u),v]$$

$$(x \in I; \ u,v \in \mathbb{R}_{++}) \ ,$$

as we have seen in section 6, the general continuous solution, nonconstant in the second variable, is given by (6.7) or, equivalently, by

$$G(x,u) = \phi^{-1}[\phi(x)u]$$

$$(x \in I, \ u \in \mathbb{R}_{++}) \ .$$

If $h(x) = G(x,\alpha)$ is given, then we have

$$\phi[h(x)] = \phi(x)\alpha$$

that is, the *Schröder equation* (7). Actually, (12) with $\phi(x) = \alpha^{\psi(x)}$ ($\alpha > 0$, $\alpha \neq 1$) can be written as

$$\Psi(x,s) = \phi^{-1}[\alpha^s \phi(x)]$$

(and that is the form in which it is often used in iteration theory). But then (13) translates into

$$h(x) = \phi^{-1}[\alpha\phi(x)] \ ,$$

or

$$\phi[h(x)] = \alpha\phi(x) \quad (x \in I) \ ,$$

again the *Schröder equation*.

In what follows, we will take in (1) and (5)

$$h(x) = x^p \quad ,$$

(p a real constant), as suggested by C. Wagner, thus getting from (1)

$$f(x_1^p,...,x_n^p) = f(x_1,...,x_n)^p \quad (x_1,...,x_n \in P) \qquad (14)$$

which is still a generalization of (7.40). We exclude $p = 1$ in which case (14) does not say anything and $p = 0$ in which case it states only $f(1,...,1) = 1$. We have taken care of the case $p = -1$ (that is, of (7.40)) in Section 7. Equation (14) with $p \neq -1,0,1$ makes practical sense for measure judgements. For instance, if the k-th individual judges the length of a side of a square to be x_k, she or he will estimate the area of the square as x_k^2 and (14) for $p = 2$ states that the synthesized judgement follows the same pattern. The role of (14) with $p = 3$ is the same for cubes. (Admittedly, this contains the assumption that the synthesizing function for areas or volumes is the same as that for lengths.)

In dealing with (14), we observe first that the root-mean-powers (7.63) do not satisfy (14) for $p \neq -1,0,1$ either except if $x_1 = ... = x_n$. So in the last part of Corollary 7.7 we may replace (7.40) by (14). Second, *we want P in (14) to be an interval, which with x contains also x^p and $x^{1/p}$. This is satisfied for $p \neq -1,0,1$ only if*

$$P = {]}0,1{[} \ or \ {]}0,1{]} \ or \ {[}1,\infty{[} \ or \ {]}1,\infty{[} \ or \ {]}0,\infty{[} = I\!R_{++}$$

$$(if \ p < 0, \ p \neq -1, \ then \ P = I\!R_{++}) \ . \qquad (15)$$

One would expect, since (7.40) implied (7.52) for the 'generating function' ψ of the quasiarithmetic mean (7.39), that (14) would imply

$$\psi(x^p) = p\psi(x) + \beta$$

or, with $\omega(t) = \psi(e^t) + \dfrac{\beta}{p-1}$,

$$\omega(pt) = p\omega(t) \ .$$

But this is not so. For the quasiarithmetic mean

$$f(x_1,...,x_n) = \psi^{-1}(\frac{1}{n} \sum_{k=1}^{n} \psi(x_k))$$

$$(x_k \in P; \ k=1,...,n) \ , \tag{16}$$

(cf. (7.39)), as seen by putting $h(x) = x^p$ into (5), equation (14) gives

$$\psi(x^p) = \alpha\psi(x) + \beta \quad (x \in P) \ , \tag{17}$$

where, as we will see, $\alpha \neq 0$ can be almost arbitrary and needs not to be p. (In (7.52), $\alpha = -1$ was a consequence of the involutory property $(x^{-1})^{-1} = x$ of x^{-1}.)

For simpler handling we again transform (17) with $x = e^t$ and with

$$\phi(t) = \psi(e^t) \quad (t \in P' = \log P)$$

into

$$\phi(pt) = \alpha\phi(t) + \beta$$

$$(t \in P') \ \text{with} \ \alpha \neq 0, \ p \neq -1,0,1 \ . \tag{18}$$

From (15) and from $P' = \log P$ we see that *we have*

$$P' = \ ]-\infty,0[\ \textit{or} \ ]-\infty,0] \ \textit{or} \ [0,\infty[\ = \mathbb{R}_+$$

$$\textit{or} \ ]0,\infty[\ = \mathbb{R}_{++} \ \textit{or} \ ]-\infty,\infty[\ = \mathbb{R}$$

$$(\textit{if} \ p < 0, \ p \neq -1, \ \textit{then} \ P' = \mathbb{R}) \ . \tag{19}$$

We are looking for strictly monotonic ψ and thus ϕ (the continuity of ψ and ϕ does not matter right now). This will give some mild restrictions on α and β in (18) or (17), some for all, others only for certain p.

Lemma 1. *The equation (18) has no strictly monotonic solution if*

(i) $\alpha = 1$ *and* $\beta = 0$,

(ii) $\alpha = 1$ *and* $p < 0$,

(iii) $\alpha < 0 < p$ *or* $\alpha > 0 > p$ $(\alpha \neq 1)$,

(iv) $\alpha = -1$.

Proof. In case (i) $\phi(pt) = \phi(t)$, so ϕ is not strictly monotonic. In case (ii), if also $\beta > 0$, then

$$\phi(pt) = \phi(t) + \beta > \phi(t)$$

and so

$$\phi(p^2t) > \phi(pt) > \phi(t) \ . \qquad\qquad (20)$$

But, since $p < 0$, we have $t > pt$ and $pt < p^2t$ for all $t > 0$ and $t < pt$, $pt > p^2t$ for all $t < 0$. Therefore (20) shows that ϕ is not strictly monotonic. The proof is similar if $\beta < 0$ (we have already taken care of the case $\beta = 0$, $\alpha = 1$ in (i)).

In case (iii), the function ω defined by

$$\omega(t) = \phi(t) + \frac{\beta}{\alpha-1}$$

$(\alpha \neq 1)$ satisfies

$$\omega(pt) = \alpha\omega(t) \ , \qquad\qquad (21)$$

and should also be strictly monotonic. If $\alpha < 0 < p$ we choose $t_o \neq 0$ so that $\omega(t_o) \neq 0$ (from (21) $\omega(0) = 0$ follows since $\alpha \neq 1$). Without loss of generality let $\omega(t_o) > 0$. Then $\omega(pt_o) = \alpha\omega(t_o) < 0$ and $\omega(p^2t_o) = \alpha^2\omega(t_o) > 0$. But these three inequalities contradict the monotonicity of ω since for $t_o > 0$ we have $t_o < pt_o < p^2t_o$ if $p > 1$ and $t_o > pt_o > p^2t_o$ if $0 < p < 1$. If $p < 0 < 1 < \alpha$ (and still $t_o > 0$, $\omega(t_o) > 0$) then $\omega(t_o) < \omega(pt_o) < \omega(p^2t_o)$ and if $p < 0 < \alpha < 1$ then $\omega(t_o) > \omega(pt_o) > \omega(p^2t_o)$ while $t_o > pt_o$ and $pt_o < p^2t_o$ so ω is again not strictly monotonic. The cases where $t_o < 0$ or $\omega(t_o) < 0$ are handled similarly.

Finally, in case (iv) applying (21) twice, we get $\omega(p^2t) = \omega(t)$ $(p \neq \pm 1)$, so ω, and thus ϕ, is not strictly monotonic either.   □

Lemma 2. *If* $0 \in P'$ *then (18) has no strictly monotonic solutions in the following cases either:*

(v) $\alpha = 1$,

(vi) $|\alpha| < 1 < |p|$ *or* $|p| < 1 < |\alpha|$.

Proof. (v) If $\alpha = 1$, put into (18) $t = 0$ and get $\phi(0) = \phi(0) + \beta$ which is impossible if $\beta \neq 0$. The case $\beta = 0$, $\alpha = 1$ was already handled in (i).

(vi) In this case $\alpha \neq 1$, so we can transform (18) again into (21) and we choose again $\omega(t_o) \neq 0$. For $|\alpha| > 1 > |p|$, $\omega(p^n t_o) = \alpha^n \omega(t_o)$ shows as $n \to \infty$, that ω tends to infinity (or oscillates when $\alpha < 0$) when the argument tends to 0 while from (21) $\omega(0) = 0$. So ψ is not strictly monotonic. The case $|\alpha| < 1 < |p|$ reduces to the case just settled if we notice that (21) implies $\omega(\frac{1}{p}t) = \frac{1}{\alpha}\omega(t)$.   □

In all other cases there exist even continuous strictly monotonic solutions. *If* $P' = \mathbb{R}_{++} = {]0,\infty[}$ *or* $P' = {]-\infty,0[}$, *then we have to have* $p > 0$ *and so, by Lemma 1, also* $\alpha > 0$.

These two cases are completely similar, we choose in the first part of the following theorem $P' = \mathbb{R}_{++}$. If $p < 0$ then, see (19), $P' = \mathbb{R}$.

Theorem 3. *With the exception of the cases listed in Lemma 1, the equation (18) with $P' = \mathbb{R}_{++}$ and $p > 0$ $(p \neq 0)$ always has a continuous strictly monotonic solution and the general strictly monotonic solution is constructed in the following way:*

(a)   *If $\alpha = 1$, $\beta \neq 0$ then $\phi(t) = \phi_o(t)$ for $t \in Q = [min(1,p), max(1,p)]$ where $\phi_o(p) = \phi_o(1) + \beta$ but otherwise ϕ_o is an arbitrary strictly monotonic function on Q and ϕ is extended to $\mathbb{R}_{++}$ by repeated application of the equations $\phi(pt) = \phi(t) + \beta$ and $\phi(t/p) = \phi(t) - \beta$.*

(b)   *If $\alpha > 0$, $\alpha \neq 1$, then $\phi(t) = \omega(t) + \dfrac{\beta}{1-\alpha}$ for all $t \in \mathbb{R}_{++}$ with $\omega(t) = \omega_o(t)$ for $t \in Q$, where $\omega_o(p) = \alpha\omega_o(1)$ but otherwise ω_o is an arbitrary strictly monotonic function on Q and ω is extended to $\mathbb{R}_{++}$ by repeated use of the equations*

$$\omega(pt) = \alpha\omega(t) \quad and \quad \omega(t/p) = \omega(t)/\alpha \ . \tag{22}$$

With the exception of the cases listed in Lemmata 1 and 2, equation (18) always has a continuous and strictly monotonic solution, also if $0 \in P'$. In these cases $\alpha \neq 1$ and we construct the general strictly monotonic solution $\phi(t) = \omega(t) + \dfrac{\beta}{1-\alpha}$ as follows. We define $\omega(0) = 0$. Furthermore:

(c)   *If P' has also negative elements (and thus, by (19), contains all negative reals) but $p > 0$ $(p \neq 1)$, then we define ω on Q as in (b) and by $\omega(t) = \omega_-(t)$ on $-Q = [min(-1,-p), max(-1,-p)]$ where ω_- is of opposite* *sign than* ω_o, *strictly monotonic on* $-Q$ *in the same sense as* ω_o *is on* Q *and such that* $\omega_-(-p) = \alpha\omega_-(-1)$ *but otherwise arbitrary and we again extend* ω *by (22).*

(d) *If* $p < 0$ *and therefore* $P' = \mathbb{R}$, *then* $\omega(t) = \tilde{\omega}(t)$ *on* $\tilde{Q} = [min(1,p^2), max(1,p^2)]$ *where* $\tilde{\omega}$ *is strictly monotonic on* $\tilde{Q}$ *with* $\tilde{\omega}(p^2) = \alpha^2\tilde{\omega}(1)$ *but otherwise arbitrary and* ω *is extended to* $\mathbb{R}$ *(both to positive and negative reals) by repeated application of (22).*

The general continuous strictly monotonic solutions are obtained by choosing ϕ_o, ω_o, ω_- *and* $\tilde{\omega}$ *in the above constructions also continuous.*

Corollary 4. *All continuous, strictly monotonic generating functions* ψ *of quasiarithmetic means (16) [or, what is the same, of the general functions (7.35) with continuous, nonconstant* g *and continuous, cancellative, associative* $\circ$ *and with (7.37)] which satisfy (14) with* $p \neq -1, 0, 1$, *where* P *is one of the intervals (15), are given by* $\psi(x) = \phi(\log x)$ *with* ϕ *as constructed in Theorem 3, where* $\alpha \neq 0$, β *are arbitrary constants with exclusion of the cases listed in Lemmata 1 and 2.*

The proofs of both Theorem 3 and Corollary 4 are obvious.

As we see, we still have many quasiarithmetic means (that is, solutions of (7.35) and (7.37)) satisfying (14). So the natural question arises, what happens if (14) is supposed for two p's, that is, for *both* p *and* $q \neq p$. Again we exclude $-1, 0$ and 1 for p and q. Furthermore, since

$$f(x_1^p, ..., x_n^p) = f(x_1, ..., x_n)^p \quad (x_1, ..., x_n \in P) \qquad (14)$$

implies

$$f(x_1^{p^m},...,x_n^{p^m}) = f(x_1,...,x_n)^{p^m}$$

for all integer m, we do not get much new information if $|p|^m = |q|^\ell$ with integer m, ℓ, that is, if $\log|p|/\log|q|$ is rational or ∞. So *we will suppose that* $\log|p|/\log|q|$ *is a finite irrational number* (as is the case in our above examples $p = 2$, $q = 3$). This takes care also of the exclusions of the exponents $-1,0$ and 1. Supplementing (14) for $p \neq -1,0,1$ with

$$f(x_1^{-1},...,x_n^{-1}) = f(x_1,...,x_n)^{-1}$$

$$(x_k \in P;\ k=1,...,n) \qquad (23)$$

would anyway restrict the general solutions given in Corollary 4 and Theorem 3 only slightly (one would have to choose ω as an odd function in (c) and (d)). But, as we will see, if added to *both* (14) *and*

$$f(x_1^q,...,x_n^q) = f(x_1,...,x_n)^q$$

$$(x_k \in p;\ k=1,...,n) \qquad (24)$$

($\log|p|/\log|q|$ is finite, irrational), it may play some role. — Of course, *P has again to be one of the intervals (15)* ($P = \mathbb{R}_{++}$ *if either $p < 0$ or $q < 0$*). We prove first the following.

Proposition 5. *If $f : P^n \to \mathbb{R}_{++}$ is continuous and satisfies both (14) and (24) where $\log|p|/\log|q|$ is a finite irrational number and*

$$P =]0,1[\ or\ ]0,1]\ or\ [1,\infty[\ or\ ]1,\infty[\ or\ ]0,\infty[\ = \mathbb{R}_{++}$$

$$(if\ p < 0\ or\ q < 0\ then\ P = \mathbb{R}_{++})\ , \qquad (25)$$

then

$$f(x_1^s,...,x_n^s) = f(x_1,...,x_n)^s$$

$$\text{whenever } s \in \mathbb{R}_{++} \text{ and } x_k, x_k^s \in P \ (k=1,...,n) \ . \quad (26)$$

If, moreover, $p < 0$ or $q < 0$ (or both) then, in addition to (26), also (23) holds, consequently (26) holds for all real s (not only for $s > 0$):

$$f(x_1^s,...,x_n^s) = f(x_1,...,x_n)^s$$

$$\text{whenever } s \in \mathbb{R}; \ x_k, x_k^s \in P \ (k=1,...,n) \ . \quad (27)$$

Proof. Repeated application of (14) and (24) gives

$$f(x_1^{p^{2m}q^{2\ell}},...,x_n^{p^{2m}q^{2\ell}}) = f(x_1,...,x_n)^{p^{2m}q^{2\ell}}$$

$$(m,\ell \in \mathbb{Z})$$

($\mathbb{Z}$ the set of all integers). Since $\{p^{2m}q^{2\ell} \,|\, m,\ell \in \mathbb{Z}\}$ is dense in $\mathbb{R}_{++}$ if $\log p^2 / \log q^2 = \log |p| / \log |q|$ is a finite irrational number (see e.g. Niven [56]) and since f is continuous, we have (26).

If, for instance, $p < 0$, then we combine (14) with (26) for $\lambda = -p$ and get (23):

$$f(x_1^{-1},...,x_n^{-1}) = [f(x_1^{-1},...,x_n^{-1})^{-p}]^{1/(-p)}$$

$$= f(x_1^p,...,x_n^p)^{-1/p} = f(x_1,...,x_n)^{p(-1/p)}$$

$$= f(x_1,...,x_n)^{-1} \ .$$

Furthermore, the combination of (23) and (26) gives (27) (also for $s = 0$ because, as we have seen, (23) with $x_1=...=x_n = 1$ gives $f(1,...,1) = 1$). $\qquad\square$

If f is a quasiarithmetic mean (16), then (26) and (27) go over into

$$\psi^{-1}[\frac{1}{n} \sum_{k=1}^{n} \psi(x_k^s)] = (\psi^{-1}[\sum_{k=1}^{n} \psi(x_k)])^s \qquad (28)$$

$$(x_k, x_k^s \in P; \ k=1,...,n)$$

for all $s \in \mathbb{R}_{++}$ or $s \in \mathbb{R}$, respectively. Writing $x_k = e^{t_k}$ $(k=1,...,n)$ and reintroducing

$$\phi(t) = \psi(e^t) \quad (t \in P' = \log P) \quad , \qquad (29)$$

(28) becomes

$$\phi^{-1}(\frac{1}{n} \sum_{k=1}^{n} \phi(st_k)) = s\phi^{-1}(\frac{1}{n} \sum_{k=1}^{n} \phi(t_k))$$

$$(t_k, st_k \in P'; \ k=1,...,n) \quad . \qquad (30)$$

If s runs through $\mathbb{R}_{++}$ (as in (26)), then this is, except for the notations, exactly (7.56) and we can copy the solution from section 7, (actually, it is somewhat easier, since P' has to be one of the intervals (19)). So we have the part (α) and essentially also (β) of the following theorem which is about (26), so *gives the solution of our problem if $p > 0$, $q > 0$ (and $\log p / \log q$ a finite irrational number).*

Theorem 6. *Let P be one of the intervals (15). Then the continuous, strictly increasing quasiarithmetic mean (16) satisfies (28) if and only if it has one of the following forms:*

(α) *For $P =]1,\infty[$ either*

$$f(x_1,...,x_n) = \exp[\frac{1}{n} \sum_{k=1}^{n} (\log x_k)^C]^{1/C} \tag{31}$$

for some constant $C \neq 0$ or

$$f(x_1,...,x_n) = \exp(\prod_{k=1}^{n} (\log x_k)^{1/n}) \tag{32}$$

(β) For $P = \,]0,1[$ either

$$f(x_1,...,x_n) = \exp(-[\frac{1}{n} \sum_{k=1}^{n} (-\log x_k)^C]^{1/C}) \tag{33}$$

($C \neq 0$) or

$$f(x_1,...,x_n) = \exp(-\prod_{k=1}^{n} (-\log x_k)^{1/n}) \ . \tag{34}$$

(γ) For $P = [1,\infty[$ only (31) with $C > 0$, for $P = \,]0,1]$ only (33) with $C > 0$.

(δ) For $P = \mathbb{R}_{++}$, in (16)

$$\psi(x) = \begin{cases} A(\log x)^C + D & \text{for } x > 1 \\ D & \text{for } x = 1 \\ B(-\log x)^C + D & \text{for } x < 1 \end{cases} \tag{35}$$

where $C > 0$, $AB < 0$, otherwise A,B,C,D are arbitrary real constants.

Proof. As mentioned above, (α) and partly (β) are consequences of Lemma 7.5. (For $P = \,]0,1[$, that is, case (β), we have $\log x < 0$ if $x \in P$, so we make the reduction to (30) using $x_k = e^{-t_k}$ and $\phi(t) = \psi(e^{-t})$ rather than (29), in order to get powers of the *positive* $(-\log x_k)$ in (33) and (34).)

If $1 \in P$ then solutions (32), (34) and (31), (33) with $C < 0$ are not continuous and strictly increasing. Indeed, in

(32) and (34)

$$f(1,x_2,...,x_n) = 1 \quad \text{for all} \quad x_2,...,x_n \in P$$

which is not *strictly* increasing, for instance in x_2. If $C < 0$ in (31) or (33), then $f(1,x_2,...,x_n)$ does not exist and $\lim_{x_1 \to 1} f(x_1,x_2,...,x_n) = 1$ is again not strictly increasing in x_2. Thus in case (γ) only (31) or (33) remain with $C > 0$.

In case (δ), both (31) has to hold for $x_k \geq 1$ with $C > 0$ and (33) for $x_k \leq 1$ with a possibly different $C' > 0$. So in (16)

$$\psi(x) = \begin{cases} A(\log x)^C + D_+ & \text{for } x > 1 \\ D_0 & \text{for } x = 1. \\ B(-\log x)^{C'} + D_- & \text{for } x < 1 \end{cases}$$

Since ψ has to be continuous, $D_+ = D_0 = D_- = D$. So ψ is indeed of the form

$$\psi(x) = \begin{cases} A(\log x)^C + D & \text{for } x > 1 \\ D & \text{for } x = 1. \\ B(-\log x)^{C'} + D & \text{for } x < 1 \end{cases} \tag{36}$$

The strict monotonicity of ψ implies that $AB < 0$. If one chooses $x_1 > 1 > x_2$ then (27) with (16) and (36) gives after an easy but tedious calculation $C' = C$. $\qquad \square$

In the cases where *at least one of p,q is negative* we have $P = \mathbb{R}_{++}$ and (27), so *the solution to our problems is contained in the following.*

Theorem 7. *Let $P = \mathbb{R}_{++}$. Then the continuous and strictly increasing quasiarithmetic mean (16) satisfies (27) if and only if*

$$\psi(x) = A \left| \log x \right|^C \text{sign} (\log x) + D$$

$$(x \in \mathbb{R}_{++}) \ , \tag{37}$$

where $C > 0$, $A \neq 0$, but otherwise A,B,C,D are arbitrary real constants.

Proof. Since (27) contains (26), we can apply Theorem 6 and since $P = \mathbb{R}_{++}$ only the solution (δ) comes into consideration. For $\lambda = -1$, (27) goes over into (23) and, as we have seen in (7.53) (and in this section before), this means (with $D = \beta/2$) that

$$\psi(x) = \omega (\log x) + D$$

where ω is an odd function. So in (36) $B = -A$ and we have (37).   □

Of course

$$\text{sign } t = \begin{cases} 1 & \text{if } t > 0 \\ 0 & \text{if } t = 0 \\ -1 & \text{if } t < 0 \end{cases} .$$

Theorems 6 and 7 furnish all synthesizing functions f, separable in the sense (7.35) (g continuous, nonconstant, $\circ$ continuous, cancellative and associative) satisfying the the unanimity condition (7.37) and (14), (24), where $\log |p| / \log |q|$ is finite and irrational.

For Lemmata 1, 2, Corollary 4, Proposition 5 and Theorems 3, 6, 7 see Aczél-Alsina [84], for a generalization (corresponding to $f(x_1,...,x_n) = g_1(x_1)...g_n(x_n)$) Aczél [84].

Only for $C = 1$ does (37) give the geometric mean.

A condition connecting measure and ratio judgements, which will yield exactly the geometric mean, is

$$R(\frac{y_1}{x_1}, \ldots, \frac{y_n}{x_n}) = \frac{f(y_1,\ldots,y_n)}{f(x_1,\ldots,x_n)}$$

$$(x_1,\ldots,x_n,y_1,\ldots,y_n \in P) \qquad (38)$$

which again may mean the following. If the k-th individual judges the measures (lengths, areas, weights, etc.) of two objects to be y_k and x_k, then he/she will presumably estimate their ratio to be y_k / x_k. If f is the synthesizing function of the measure judgements and R that of the ratio-judgements, then (38) states that the synthesized ratio-judgement should also be the quotient of the synthesized measure judgements (Aczé l-Alsina [86]).

We can arrive at (38) also in another way which establishes connections with economics on the one hand and with the most important (or traditional) 'law of science', namely (4.7), on the other (thus closing another 'circle'). Suppose that the k-th individual estimates today's price level as x_k, that of the next year as y_k, then she/he estimates the inflation rate to be $r_k = y_k/x_k$. So $y_k = r_k x_k$ $(k=1,\ldots,n)$. If f is the synthesizing function of the price level estimations and R that of the inflation rate judgements, then the supposition that these behave in the same way is expressed by the equation

$$f(r_1 x_1,...,r_n x_n) = R(r_1,...,r_n)f(x_1,...,x_n)$$

$$(x_1,...,x_n,r_1 x_1,...,r_n x_n \in P) \ . \quad (39)$$

This is the same as (38), but also as (4.7), except that $x_k, r_k x_k$ $(k=1,...,n)$ are restricted to the subinterval P of positive numbers. But, applying the analogue of Corollary 5.5 to n-place functions, we see that (39) (and thus (38)) can be extended so that $I\!R_{++}$ stands in place of P and from Corollary 4.3 we get *without any separability* (and in the first part also without unanimity) *condition* the following.

Corollary 8. *The general positive solutions of (38) (where P is an interval of positive numbers), with, say, f bounded from above on an open n-dimensional interval, are given by*

$$f(x_1,...,x_n) = a \prod_{k=1}^{n} x_k^{q_k} \quad (x_1,...,x_n \in P) \qquad (40)$$

and

$$R(r_1,...,r_n) = \prod_{k=1}^{n} r_k^{q_k}$$

$$(r_1,...,r_n \in P/P = \{y/x \mid x \in P, y \in P\}) \qquad (41)$$

where $a > 0$ but otherwise $a, q_1,...,q_n$ are arbitrary constants.

If also the unanimity condition $f(x,...,x) = x$ $(x \in P)$ is supposed, then $a = 1$ and $\sum_{k=1}^{n} q_k = 1$. If the boundedness condition is replaced by strict increasing, say of f, at least at one point for each variable, then $q_k > 0$ $(k=1,...,n)$.

So this time *we have characterized exactly the weighted geometric means* $\prod_{k=1}^{n} x_k^{q_k}$ $(\sum_{k=1}^{n} q_k = 1, \ q_k > 0; \ k=1,...,n)$ *as synthesizing functions. If symmetry is also supposed, we get the ordinary geometric mean (7.64).*

After extending (38) to all positive reals and putting $x_1 = ... = x_n = 1$, we see that $f(y_1,...,y_n) = aR(y_1,...,y_n)$ (cf. (40) and (41)) and get

$$R(\frac{y_1}{x_1}, \ldots, \frac{y_n}{x_n}) = \frac{R(y_1,...,y_n)}{R(x_1,...,x_n)}$$

and $R(1,...,1) = 1$.

With $y_k = 1$ or $y_k = x_k^2$ $(k=1,...,n)$ this goes over into (14) with $p = -1$ (reciprocal condition) and $p = 2$ (example above), respectively. Repeated use gives (14) first for all integer, then for all rational p and, if R (or f) is also continuous, then for all real p (cf. (27)). This shows how 'powerful' the condition (38) is and explains why it gives the geometric mean so easily. (Actually, comparison of Theorem 7 and Corollary 8 shows that even (27) and (7.35) together are weaker than (38)).

In (38) we allowed *different synthesizing functions for the measures and the ratios.* We may even allow *different synthesizing functions for the x_k's and y_k's, thus replacing (38) by*

$$R(\frac{y_1}{x_1}, \ldots, \frac{y_n}{x_n}) = \frac{F(y_1,...,y_n)}{f(x_1,...,x_n)}$$

$$(x_1,...,x_n,y_1,...,y_n \in P) \ . \tag{42}$$

By Corollaries 5.5 and 4.3 we get

$$f(x_1,...,x_n) = a \prod_{k=1}^{n} x_k^{q_k},$$

$$F(y_1,...,y_n) = A \prod_{k=1}^{n} y_k^{q_k}$$

and

$$R(r_1,...,r_n) = \frac{A}{a} \prod_{k=1}^{n} r_k^{q_k}$$

as general positive solutions of (42) with one of f, F or R bounded from above. Here $a > 0$, $A > 0$ but otherwise a, A, $q_1,...,q_n$ are arbitrary constants. *If the unanimity condition holds for at least two of f, F and R, then* $a = A = \sum_{t=1}^{n} q_k = 1$ *and if the boundedness condition is replaced by strict increasing at least at one point, then* $q_k > 0$ $(k=1,...,n)$, *so we get the same weighted geometric mean for all three synthesizing functions f, F and R.*

References

(This bibliography is kept fairly short. For further functional equations in a single variable see Kuczma [68], for the Cauchy and Jensen equations (and inequality) and generalizations Kuczma [85], for equations containing several variables, with strong emphasis on applications, Aczél [66] and Aczél-Dhombres [87], for applications in economics Eichhorn [78], for applications to information measures Aczél-Daróczy [75], to the theory of measurement e.g. Krantz-Luce-Suppes-Tversky [71], to probability distributions for instance Galambos-Kotz [78], to probabilistic metric spaces Schweizer-Sklar [82]. There are many further references in these books).

Aczél, J. [66], Lectures on Functional Equations and Their Applications. Academic Press, New York-London, 1966.

Aczél, J. [83], Diamonds are not the Cauchy extensionists' best friend. C.R. Math. Rep. Acad. Sci. Canada 5 (1983), 259-264.

Aczél, J. [84], On weighted synthesis of judgements. Aequationes Math. 27 (1984), 288-307.

Aczél, J. and Alsina, C. [84], Characterizations of some classes of quasilinear functions with applications to triangular norms and to synthesizing judgements. Methods Oper. Res. 48 (1984), 3-22.

Aczél, J. and Alsina, C. [86], On the synthesis of judgements. To appear in Socio-Econom. Planning Sci.

Aczél, J., Baker, J., Djokovic´, D.Z., Kannappan, Pl. and Rado´, F. [71], Extensions of certain homomorphisms of semigroups to homomorphisms of groups. Aequationes Math. 6 (1971), 263-271.

Aczél, J. and Daróczy, Z. [75], On Measures of Information and Their Characterizations. Academic Press, New York-San Francisco-London, 1975.

Aczél, J. and Dhombres, J. [87], Functional Equations Containing Several Variables. Cambridge University Press, Cambridge, 1987.

Aczél, J. and Gehrig, W. [86], Determination of all generalized Hicks-neutral production functions. To appear.

Aczél, J., Kannappan, Pl., Ng, C.T. and Wagner, C. [83], Functional equations and inequalities in 'rational group decision making': In General Inequalities 3 (Proc. 3rd Internat. Conf. Gen. Inequ. Oberwolfach, 1981). Birkhäuser, Basel-Boston-Stuttgart, 1983, pp. 239-245.

Aczél, J., Ng, C.T. and Wagner, C. [84], Aggregation theorems for allocation problems. SIAM J. Algebraic Discrete Methods 5 (1984), 1-8.

Aczél, J., Roberts, F.S., and Rosenbaum, Z. [86], On scientific laws without dimensional constants. To appear in J. Math. Anal. Appl.

Aczél, J. and Saaty, T.L. [83], Procedures for synthesizing ratio judgements. J. Math. Psych. 27 (1983), 93-102.

Aczél, J. and Wagner, C. [81], Rational decision making generalized. C.R. Math. Rep. Acad. Sci. Canada 3 (1981), 138-142.

Bossert, W. and Pfingsten, A. [86], A new concept of inequality-equivalence. To appear.

Courant, R. and John, F. [74], Introduction to Calculus and Analysis. Vol. II. John Wiley & Sons, New York, 1974.

Daróczy, Z. and Losonczi, L. [67], Ueber die Erweiterung der auf einer Punktmenge additiven Funktionen. Publ. Math. Debrecen 14 (1967), 239-245.

Dhombres, J. and Ger, R. [78], Conditional Cauchy equations. Glas. Mat. Ser. III 13 (33), (1978), 39-62.

Eichhorn, W. [78], Functional Equations in Economics. Addison-Wesley, Reading, Mass. 1978.

Eichhorn, W. and Gehrig, W. [82], Measurement of inequality in economics. In Modern Applied Mathematics — Optimization and Operations Research. North Holland, Amsterdam-New York, 1982, pp. 657-693.

Eichhorn, W. and Voeller, J. [76], Theory of the Price Index. Fischer's Test Approach and Generalizations. Springer, Berlin-Heidelberg-New York, 1976.

Ellis, B. [66], Basic Concepts of Measurement. Cambridge University Press, London, 1966.

Förg-Rob, W. [86], Differential solutions of the translation equation. To appear.

Galambos, J. and Kotz, S. [78], Characterizations of Probability Distributions. A Unified Approach with an Emphasis on Exponential and Related Models. Springer, Berlin-Heidelberg-New York, 1978.

Gehrig, W. [76], Neutraler technischer Fortschritt und Produktionsfunktionen mit beliebig vielen Produktionsfaktoren. Hain, Meisenheim am Glan, 1976.

Hamel, G. [05], Eine Basis aller Zahlen und die unstetigen Lösungen der Funktionalgleichung $f(x+y) = f(x) + f(y)$. Math. Ann. 60 (1905), 459-462.

Hardy, G.H., Littlewood, J.E. and Pólya, G. [52], Inequalities, 2nd edition. Cambridge University Press, Cambridge, 1952.

Hicks, J.R. [32], The Theory of Wages. MacMillan, London, 1932.

Kolm, S.C. [76], Unequal inequalities, I. J. Econom. Theory 12 (1976), 416-442.

Krantz, D.H., Luce, R.D., Suppes, R. and Tversky, A. [71], Foundations of Measurement. Vol. I, Additive and Polynomial Representations. Academic Press, New York-London, 1971.

Kuczma, M. [68], Functional Equations in a Single Variable. Polish Scientific Publishers, Warszawa, 1968.

Kuczma, M. [72], Notes on additive functions of several variables. Uniw. Sląski w Katowicach Prace Nauk - Prace Mat. 2 (1972), 49-51.

Kuczma, M. [85], An Introduction to the Theory of Functional Equations and Inequalities. Cauchy's Equation and Jensen's Inequality. P.W.N.—Uniw. Sląski, Warszawa-Kraków-Katowice, 1985.

Luce, R.D. [59], On the possible psychophysical laws. Psychol. Rev. 66 (1959), 81-95.

Luce, R.D. [64], A generalization of a theorem on dimensional analysis. J. Math. Psych. 1 (1964), 278-24.

Marsden, J.E. [74], Elementary Classical Analysis. Freeman, San Francisco, 1974.

Mohr, E. [51], Bemerkungen zur Dimensionsanalysis. Math. Nachr. 6 (1951), 145-153.

Ng, C.T. [74], Representation for measures of information with the branching property. Inform. and Control 25 (1974), 45-56.

Niven, I. [56], Irrational Numbers. Mathematical Association of America — John Wiley & Sons, New York, 1956.

Ostrowski, A. [29], Mathematische Miszellen XIV. Ueber die Funktionalgleichung der Exponentialfunktion und verwandte Funktionalgleichungen. Jber. Deutsch. Math.-Verein. 38 (1929), 54-62.

Radó, F. and Baker, J. [86], Pexider's equation and aggregations of allocations. To appear.

Robinson, J. [37], Essays in the Theory of Employment. MacMillan, London, 1937.

Saaty, T.L. [80], The Analytic Hierarchy Process. Planning, Priority Setting, Resource Allocation. McGraw-Hill, New York, 1980.

Sato, R. and Beckmann, M. [68], Neutral inventions and production functions. Rev. Econom. Stu. 35 (1968), 57-66.

Schweizer, B. and Sklar, A. [82], Probabilistic Metric Spaces. North Holland, New York-Amsterdam-Oxford, 1982.

Sibirsky, K.S. [75], Introduction to Topological Dynamics. Noordhoff, Leiden, 1975. (Russian original: Akad. Nauk. Moldav. SSR, Kishinev, 1970).

Stehling, F. [74], Neutral inventions and CES production functions. In Production Theory (Proc. Internat. Sem. Univ. Karlsruhe, 1973). Springer, Berlin-Heidelberg-New York, 1974, pp. 65-94.

Stehling, F. [75], Eine neue Charakterisierung der CD- und ACMS-Produktionsfunnktionen. Methods Oper. Res. 21 (1975), 222-238.

Steinhaus, H. [20], Sur les distances des points des ensembles de mesure positive. Fund. Math. 1 (1920), 125-129.

Szymiczek, K. [73], Note on semigroup homomorphisms. Uniw. Sląski w Katowicach Prace Nauk. - Prace Mat. 3 (1973), 75-78.

Young, H.P. [86], Progressive taxation and the equal sacrifice principle. To appear.

Subject Index

Index of names
(numbers greater than 154 refer to References)